SECRETS OF THE WINDS

Discovery Chronicles of Earth's Galactic Odyssey

JOHNNY TESONE

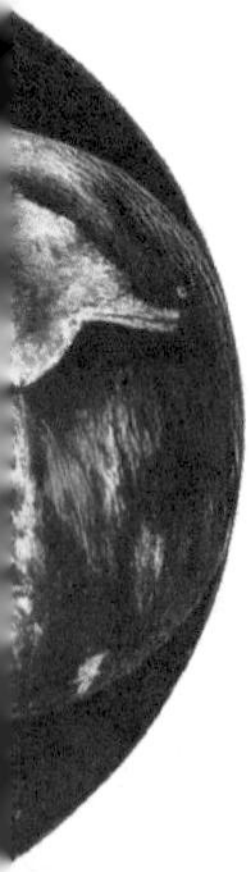

First Edition 2025

TABLE OF CONTENTS

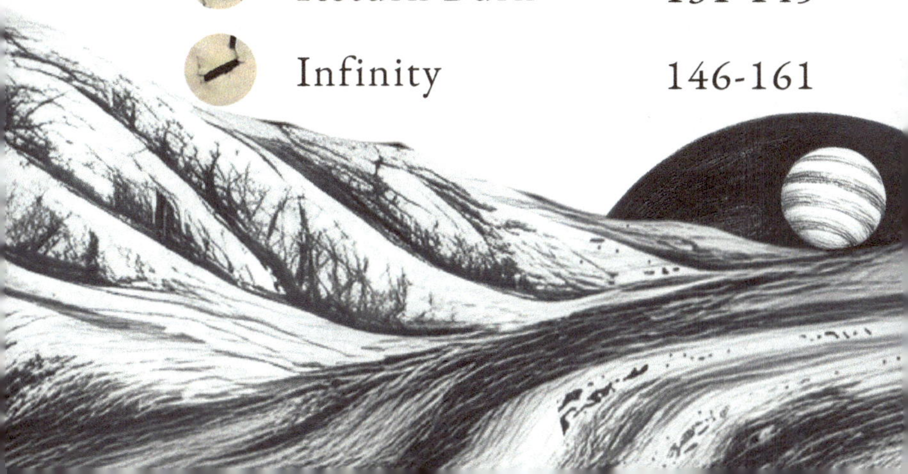

THE WINDS

There's been this powerful affinity in my life for the majestic Wind River Range of Wyoming, the Winds. This is where I believe it all started in another past life where I was an indigenous man who raised a family in the lush Wind River Valley, a land freshened with mountain water, rich in upland and big game, adorn in bountiful settings, raised on the confluence of the North and South Forks. I believe that I was first born in this land, which is now the Wind River Indian Reservation. Back in that time, I led a small tribe where I scouted, fished, hunted, made tools and weapons, farmed, created art, and wrote about various tribal scenes, many with animals and fish. This was the age of man, 10,000 years ago, a time when the Earth ice caps were melting, nomads roamed the continents and life on the planet was flourishing. I had vivid dreams of this fantastical yet primitive past life until I was six years old and probably why I've been so personally impacted in this life by indigenous people and experiences on sacred lands.

I often recall one of my favorite field trips which was in 3rd grade to South Dakota when we visited our pen pals on the Sioux Nation Tribal lands.

This time at UW the geology and anthropology curriculums were regionally and world-renowned, strongly considered premier science programs and few of my professors were the earliest pioneers in their respective fields and published many papers about WY geology, the Wind River Range along with authoring many anthropology books. This environment allured me deeper to nature and science discovery intellect and personified into a trail blazer. I knew this was the perfect college for me to learn and grow with a new lease on life without football when I had to quit the program my junior year from one too many concussions, or rather 'conclusions'. To keep my scholarship per the team and NCAA rules, I had to tutor the freshman football players in their 'Rock for Jocks classes as I soon forgot about the scolding, hyper-intense coaches, load blocking 250 lb. BYU linebackers in the Wish-Bone offense, the ice baths and the joint surgeries.

Often in practice, I would gaze at the Snowy Range to west of Laramie, dreaming away the agony by revisiting past summer days exploring the trout filled lakes for distracting me from the harsh physical reality on the grid iron.

Out of all the geographic basins and ranges that I studied in the Rocky Mtns., it was the Wind River Range that was the most enchanting, wild and the focus for me to study starting in 1981. During college, I became fascinated with the stories told by a Shoshone teammate friend, sharing his youth growing up the in the Wind River valley. I'm grateful for his native spirit, his knowledge of the best backcountry guides to waters that hold the biggest trout.

Moreso, later in our lives, it's the connective stories such as the one about his grandma warning him about the Nimerigar, "Little People' scaring her at Dinwoody Lake as they did to me in Oct. 2020.

I survived in that previous nomadic life like this one with an open attitude and aptitude for learning, having a sixth sense of desire fueled by an untapped reservoir of curiosity and creativity but powered by perseverance. I strive for mental toughness, to be a clear visionary, a true problem solver, and an explorer of the Earth and the cosmos.

The time has come for this story, crafted by a father, a geologist, fly fisherman and master fly-tier time traveler who can tell the world about the greatest Earth story that began in the Winds.

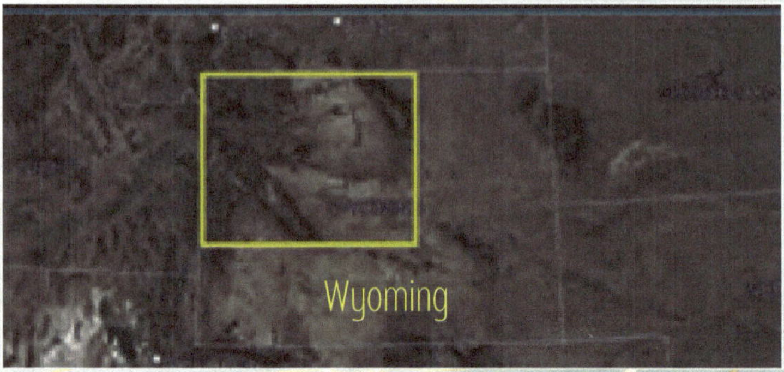

REGIONAL, AREA, WIND RIVER

WY REFERENCE MAPS

RANGE AND YELLOWSTONE

The Wind River Range trout fishing has been a personal pursuit for 30 years due to the epic "Hopper Heaven" fly-fishing days and the breathtaking mountain scenery. But the elements, bears, bugs, and snakes make the Winds way too primitive a wilderness for the unprepared. It's usually after the 3rd or 4th day that I feel I could stay in camp forever.

The thrill of fishing in the Winds has released the ultimate freedom, knowing that I've found my pure passion in a sport that has profoundly shaped my human sense of a metaphoric and transformational life. The yearly trips to the Winds have defined my foundational qualities as an explorer, geologist, artist, and author.

I recall one fall day on the Wind River when my brother Tim frustratingly asked me, "How can you catch so many?"

"Well, Tim, I've been fishing here for hundreds of years."

The Winds are most enjoyed with dear friends and family who share their best stories on the river or in camp and who have graced my life on earth.

Mike, is my best friend and long time fishing buddy. Bob, my Shoshone brethren and UW college friend. Brother Tom (RIP, center). Darwin, the renowned Shoshone backcountry guide; Carrie and Sterling, my lifelong fishing camp friends.

Thanks to my life friend, Doug Thompson, who always believed in my journey and inspired me to write about the winds.

MY DAUGHTER'S ARTWORK EXUDES GRACE, BEAUTY ,CREATIVITY, AND FREEDOM.

ARTISTS; ELLA TESONE, GIANNA TESONE

The Lewis and Clark Expedition, a monumental journey of discovery and endurance, is a testament to the human spirit's quest for knowledge and connection. It was a venture into the unknown, a daring push into the vast wilderness that defined the Western United States. The expedition's success hinged not just on the leadership of Meriwether Lewis and William Clark but also on the contributions of the diverse crew, including the invaluable Sacagawea. Her story is one of courage and resilience. She is a young Shoshone woman who, with her infant son strapped to her back, navigated the treacherous terrains and bridged the cultural divides between her people and the explorers.

Edgar Samuel Paxson: Lewis and Clark at Three Forks

William Clark (left) and Meriwether Lewis (center) with Sacagawea, detail from Lewis and Clark at Three Forks, oil painting by Edgar Samuel Paxson, 1912.History and Art Collection/Alamy.

Firing rifles as they galop in a great processional circlearound their base camp, 2000 eastern Shoshonis theartist Millers estimated 1837.

Jubilant at the arrival of summer in her Wind Riverdomain, a Shoshoni girl swings from a birch branchas a demure friend looks on.

Savoring a moment of solitude at the edge of camp ,a warrior smokes his pipe, " the great solace of hisleisure hours," said artist Miller .

WILSON CREEK BRIDGE

2004 was a heavy snowpack year in the Winds, but we already had lined up Darwin's crew weeks ahead to take us into the backcountry toa remote river canyon. The weather started turning bad the morning of the trip, so we decided instead to head to another drainage, presumably less strenuous and precarious. As we traversed Windy Gap Ridge, the weather cleared so we could see the snowpack, and then I heard our guide say, "I sure hope we can get over Wilson Creek bridge."

When we reached the bridge, Wilson Creek was roaring out of its banks. The two lead pack horses, tethered, started rearing while the first pack horse was being pulled across forcibly. But that second horse still refused to get his ass over that bridge. By this time, Mike and I dismounted, and all we could do was watch in total disbelief and fear as it stumbled and fell off the bridge, almost dragging the first horse down off the bridge with it, too!

I gasped to yell, "Cut the rope, cut the rope!" The lead horse was freed, and a second pack horse was swept downstream, submerged, groaning, eyes bulging as it gained footing, but its chest was gashed from the log downstream.

An hour of sunshine later was barely enough time to dry off what gear we could salvage to survive the elements of a hail storm lasting 4 of the first 5 days.

One day of fishing was all we had, but the lunker cutthroats in Bear Lake always hit my #18 triple hackle Renegade.

Sadly, for the horse named Idiot, it was his last trip into the Winds.

After all the episodic days, It was July 21, 2020, the day the Earth stood still, for just for me.

It was this day that the wind, mountains, water synchronized around me and created the Geovortex. This was exact time and place of the elemental earth connection to my past nomadic homesteads in the Wind Rivers.

The magical totem figurines couldn't be replicated on Google Earth, but I scoped these disk-shaped concaved outcrops distinctly patterned unlike anything I had ever seen, read or studied about in the Winds. The UFOs - Unidentified Fossilized Objects came in 1000', 23° up-dip from the Totem Rock outcrop.

STRUCTURAL 'ROCKET' GEOLOGY, WIND RIVER RANGE

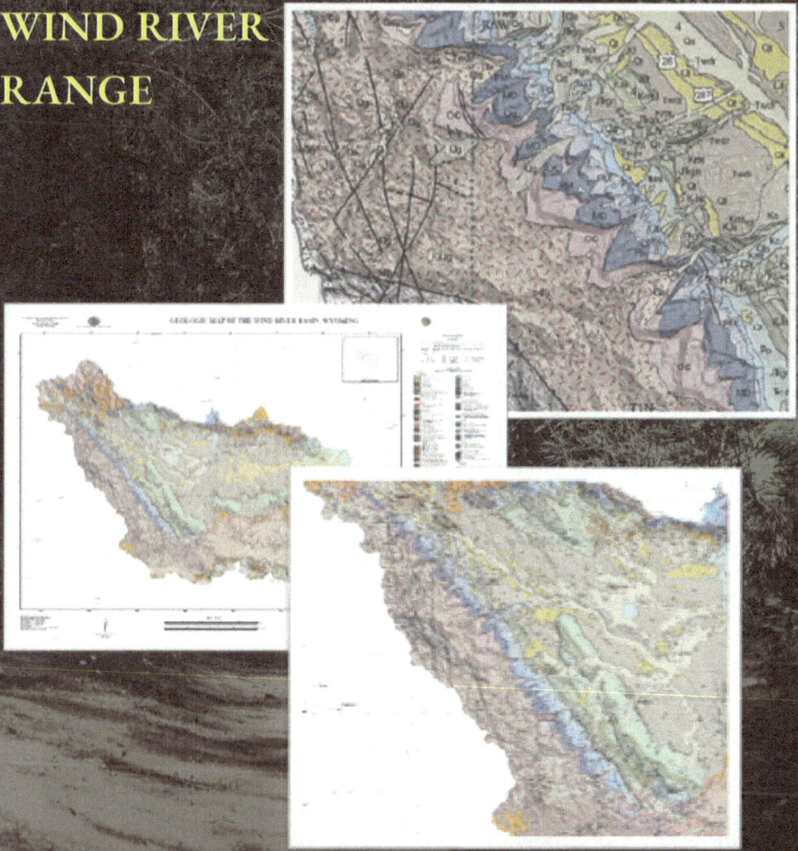

Entire Wind River Range is an angular conformity, unzippering, blasted out from main Yellowstone event to the north, extended laterally SE to NW 80 miles. Multiple, angle fold axis are mapped from Precambrian through Paleozoic, Mesozoic to the Cenozoic.

Credit; Geological Map of the Wind River Basin, WY State Geological Survey.

Structured Rocket Geology, Wind River Range

The sub rocket outline seen as fold lines depicts countless formations of a fractal nature within the focus area. The WRR itself has spawned an infinite number of rockets and is a core geological model construct. First found in WRR, the geometry correlates to global attributes.

Intersecting fold lines outline the staged geo-symmetric rocket orogeny replicated in the range. The sub-events sit on massive foundational Precambrian Rocket base outcrops. Events occurred instantly in one age but were mapped through time, and the species leaves an infinite legacy of geologic wonderment.

40-mile X-section of rockets mapped unfolding, multiple sub-processes from the Great Yellowstone Explosion.

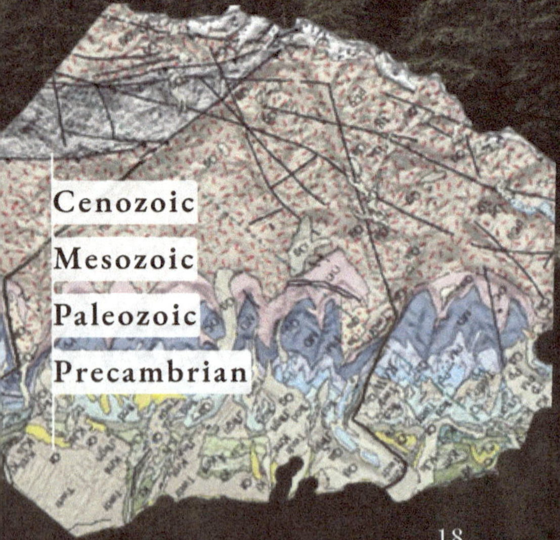

WIND RIVER ROCKET RANGE

Yellowstone explosion correlation features trace the event layers depicting land expansion and replication over time.

NW Territory Great Bear Lake

Columbia, SA - Laguna De Tota

Peru, Bolivia; Lake Titicaca

WYO - Yellowstone Lake

Western Columbia, SA

Central, WYO

9-nine-fold signature of orogenic rocket species' causing geo-cyclical events from the origin event in the Precambrian.

Multi-colored, thick iron-red layers of extreme folds and horned anchors. This is another distinct sub-process stage of the explosion result. More on the colossal Yellowstone event in the Plan chapter.

Rocket Species microscopy, Wind River Range

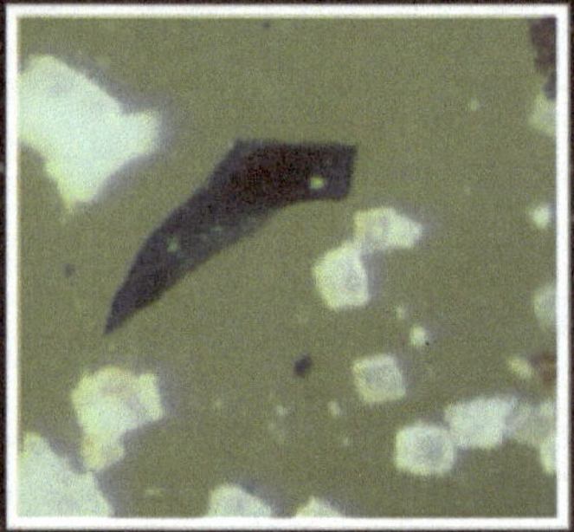

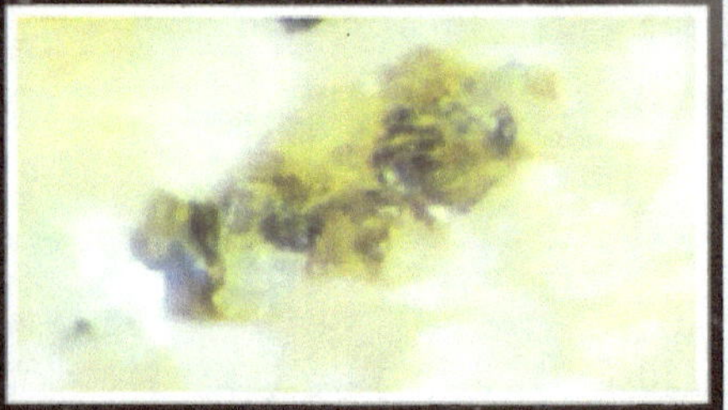

Remote canyon trails intersect the Precambrian formations, but access around the impact sites are littered with exobiological, astro geological markers. The microscopic work went into full speed with limitless samples that proved substantial to indexing new life in a mass ocean event.

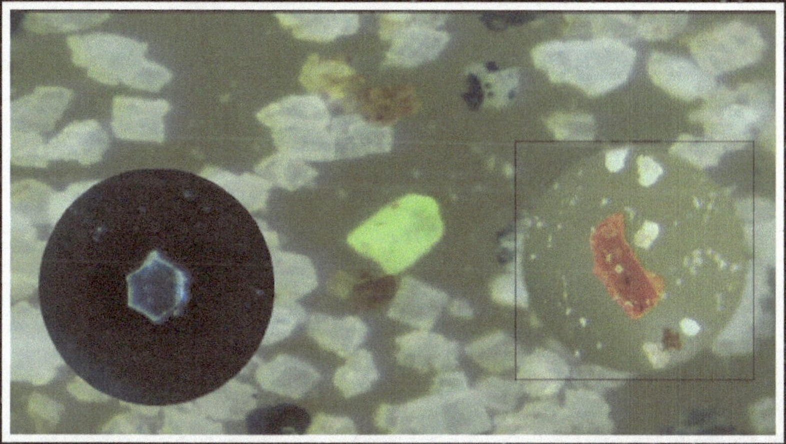

I knew early that this was not typical carbonate mineralization of dolomite, limestone found in younger aged ocean sediments recorded in drill cuttings and cores in the Rockies. After hours examining the foreign crystals, it became ana we-inspiring feeling of forever and weightlessness discovering and holding fractal alien life. It was clear to me to me then that this needed to be shared with the world and this is only the first chapter.

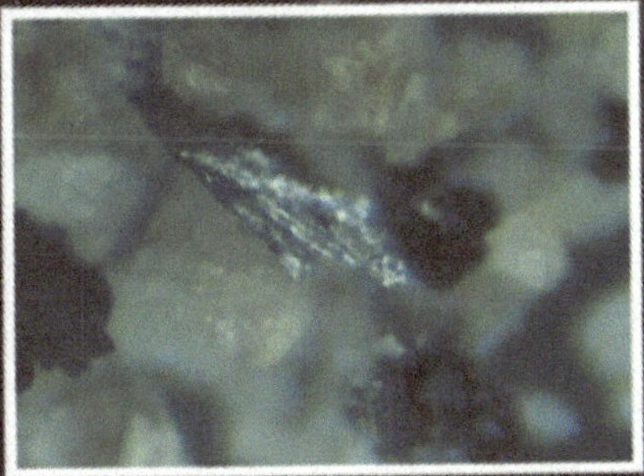

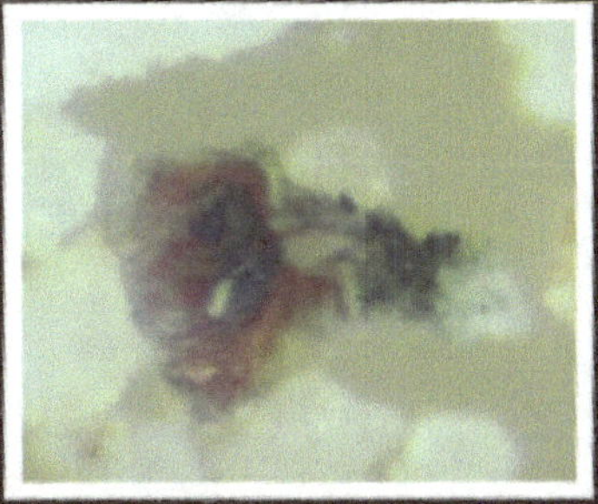

Precambrian event reveals limitless forms of organic, alien mineralization, active, fractal life, a true scientific phenomenon and a historic place holder.

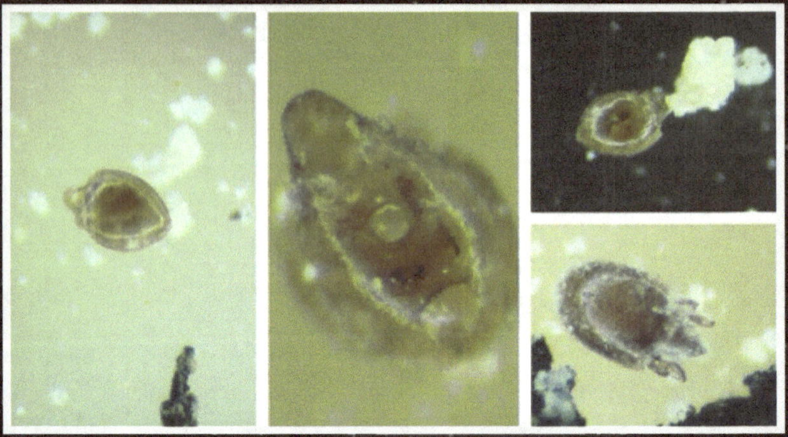

It took hundreds of hours to capture details about the Precambrian alien eggs as they were almost impossible to re-find in another microscope session.

The transparency was a view into the orifices, limbs, circuitry, and validation of the water cell battery, as seen in the middle right picture.

Sample of black translucent EIN eggs like in shape and internal electrodes. Eggs found in ant hill, and which are much larger than the WY specimens.

472-gram (1lb.) baseball size, Precambrian alien fetus egg fossil. Flat bottom, fingers poking up, out, and black oxygenation nodules are a few key attributes.

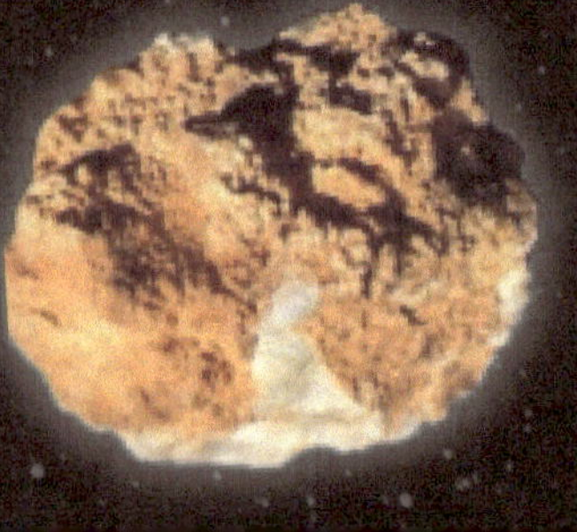

GROUND ZERO

Utilizing Google Earth over the years, generally in work projects, mapping slide areas and for new fishing trails, the tool's high-resolution capabilities opened my mind to this whole new world of Wyoming, Earth, and Astrogeology – Ground Zero.

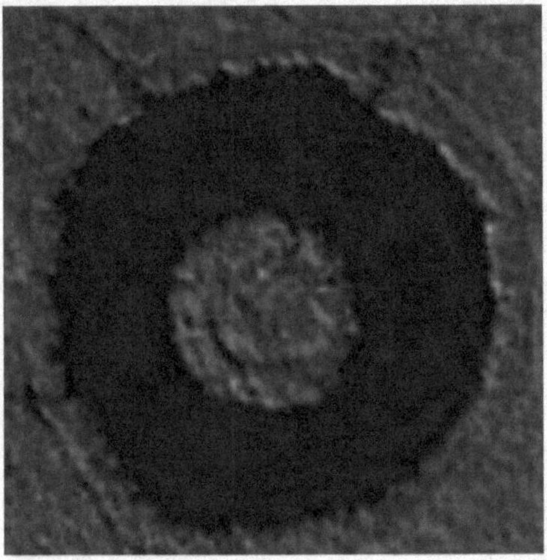

Returning to the Winds Fall 2020, I covered more expanse of the basin range surface miles east into the plains to find the scale of the events that the geology was beginning to divulge to me beyond the totem cliffs. As I explored the hills and plains, it was a time when the focus was on the larger 200' concentric, geometric outcrops and surface patterns, UFOs: Unidentified Fossilized Objects.

From the microscope to the satellite, the outer space features scaled exponentially as my knowledge accelerated beyond imagination.

The chapter highlights the Eerie Dunwoody Lakes trip from the northern Winds and the petroglyph discovery correlation to the alien embryo fossil.

Public areas exposed 40'-300' impact events: landings, sprawling, damming with mineral matrix and active alien mounds and burrows.

Visibly recent bombs pitted the paleo Wyoming surface, which must have had favorable landing conditions due to the melting of glaciers or shallow ocean conditions.

This latest bombardment event in the Holocene is the most plausible reason Earth's largest land mammals, such as the Wooly Mammoth, Giant Sloth and Sabre Tooth tiger, weren't extinct as a food source.

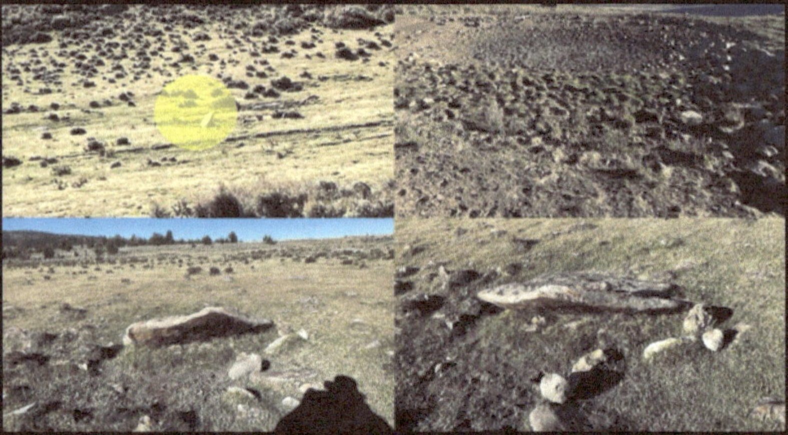

Surface Impacts that revealed scattered convex shaped, broken outer metallic heat shell built to blast through any planet atmosphere. With the right water volume balance, the species becomes a living earth system.

Millions of tons of stacked, formed rocks of various colors, brecciated are the matrix boulders of common size, with flattened facies that constructed this 300' landing and damming structure. This impact feature was the first major clue of the harvesting and reconstituting nature, evidence explaining the missing geology from the rock records.

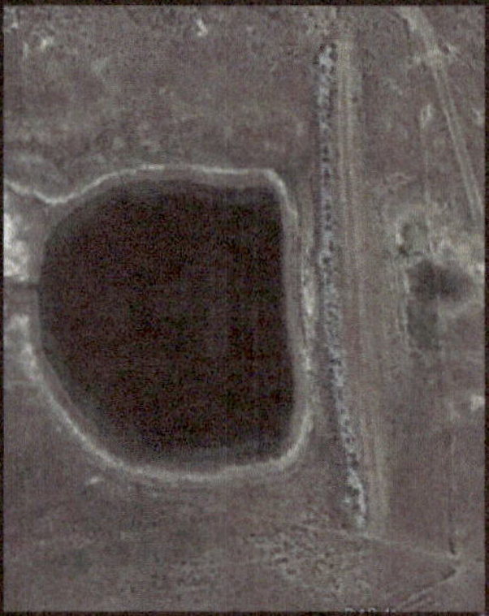

The missing Earth matter has been harvested, then remixed, reformed, and designed to build and launch the impactors.

The 10lb red specimen is ultra hard, heavy, and iron alloy-based, making the outer shell dam that absorbs and reacts with H20 through geological time. Standing samples hold water for over 3 weeks. This is the Cosmo to Earth geological continuum resource and process.

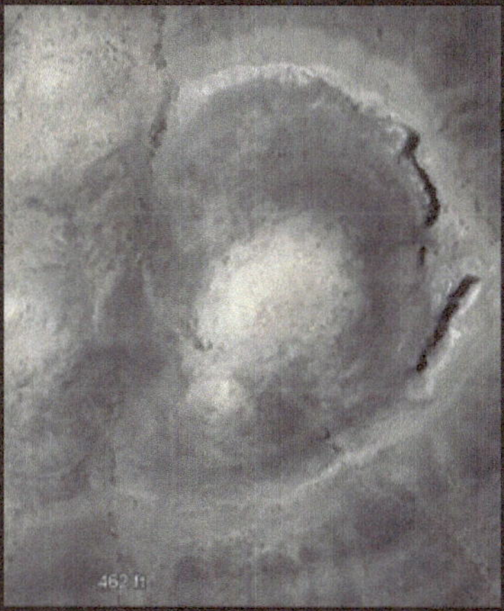

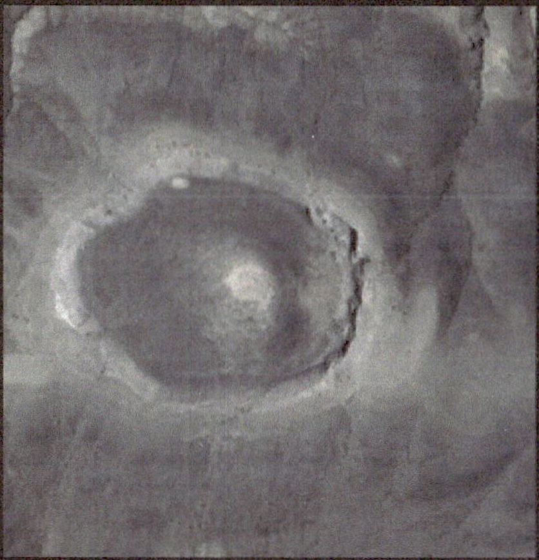

After visiting the craters, the search led to the WY plains, searching for disks, concentric surface features and if the event could be traced to Wind River Range outcrops. However, I didn't know how they were connected or if I could fully understand these forms during Ground Zero scenes.

At this time of the research, I didn't understand the geometric folds, nor the complexities of the re-conforming properties or the species.

The volume of perfect circles is not irrigated farm fields in the middle of nowhere, WY, but the bomb pipe object remnants in the earth rebuilding, reshaping, and casting mold for future combustible propulsion events.

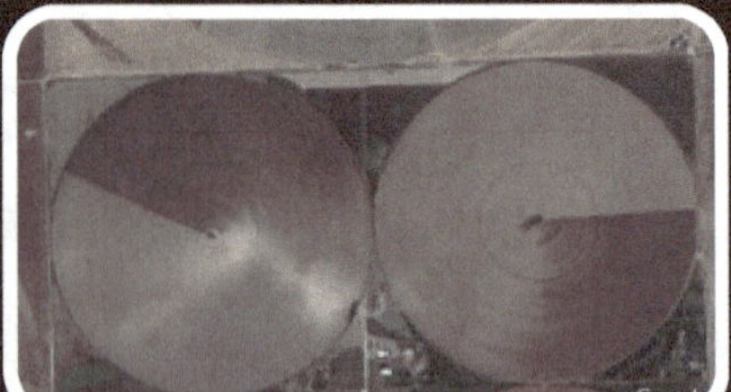

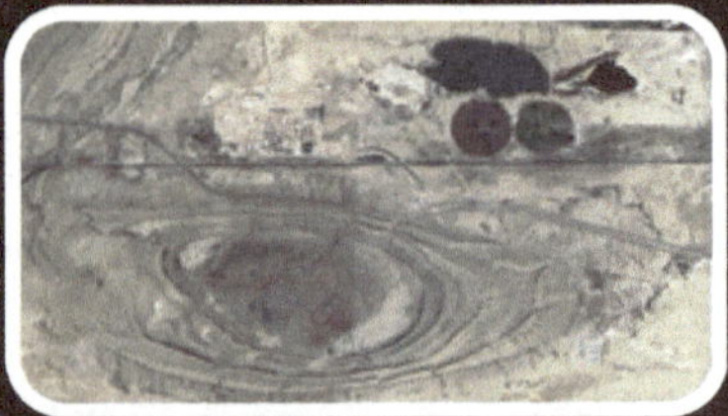

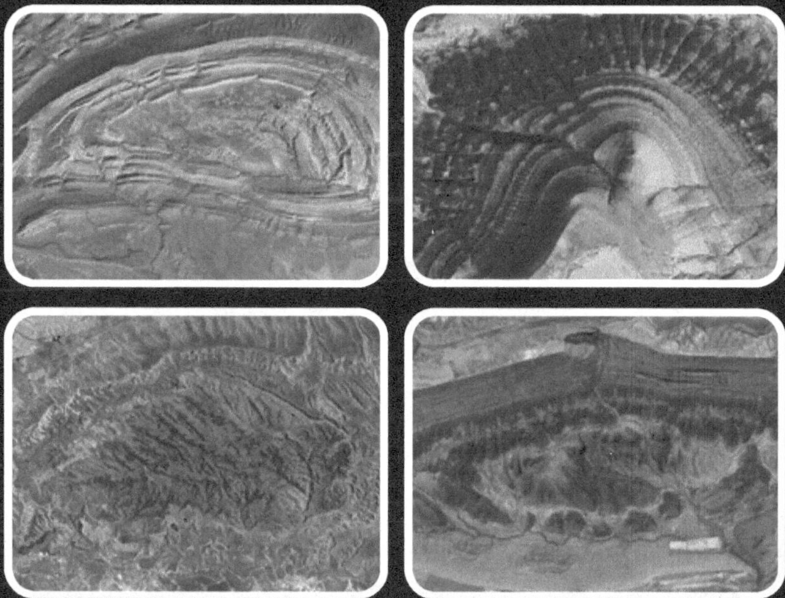

Large folds objects was huge clue of extraterrestrial organic, fractal species.

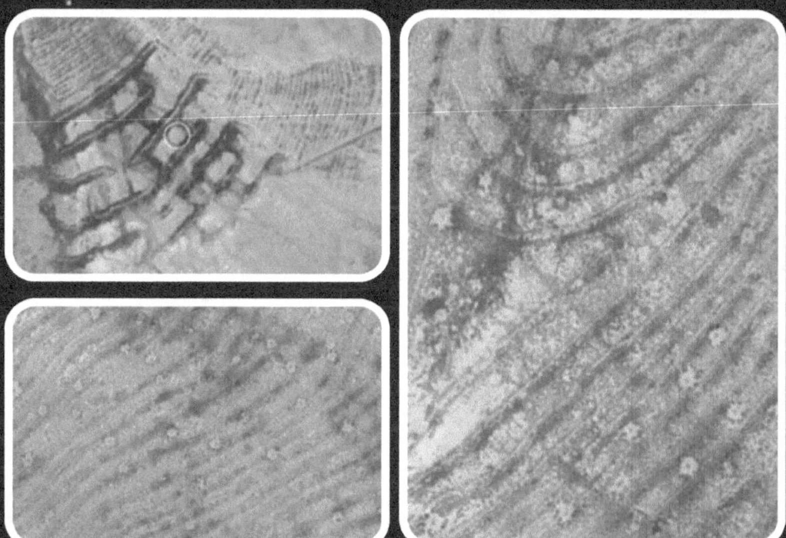

Alien geomorphology exhibits distinct, expansive and abundant surface patterns.

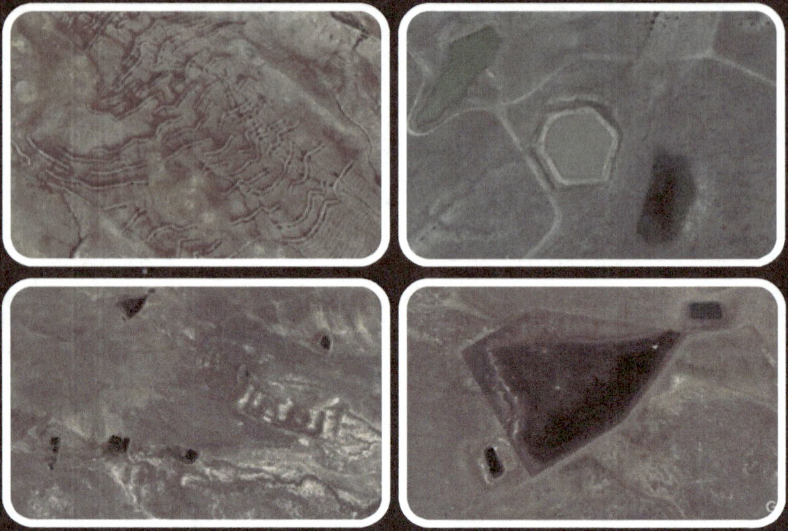

Extremely defined surface features on the WY plains reveal the early clues to processing surface for water control, elemental leaching for the species alloying propagation.

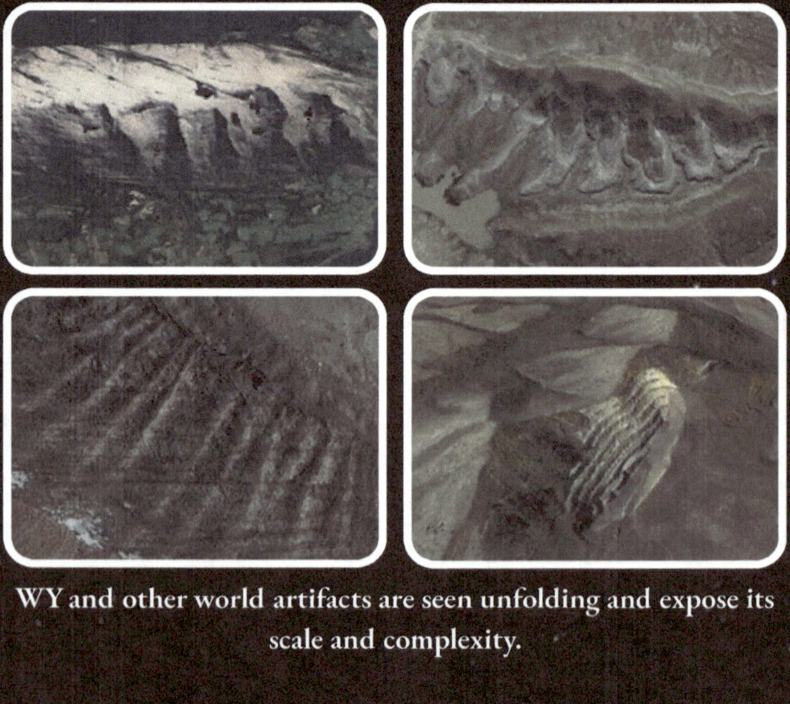

WY and other world artifacts are seen unfolding and expose its scale and complexity.

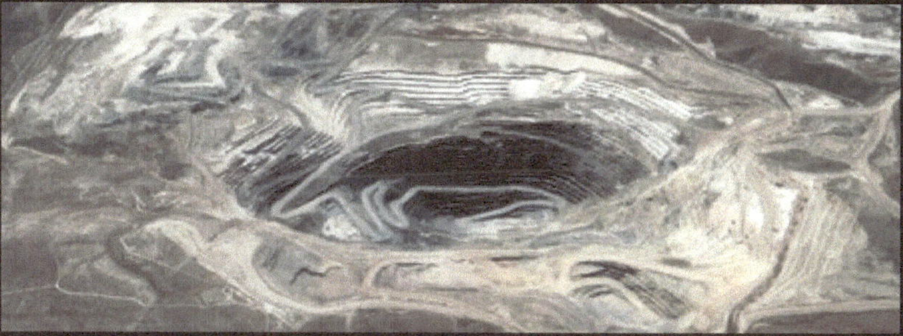

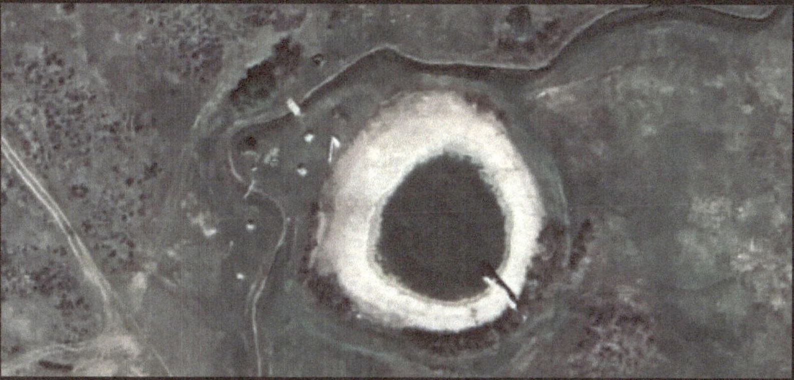

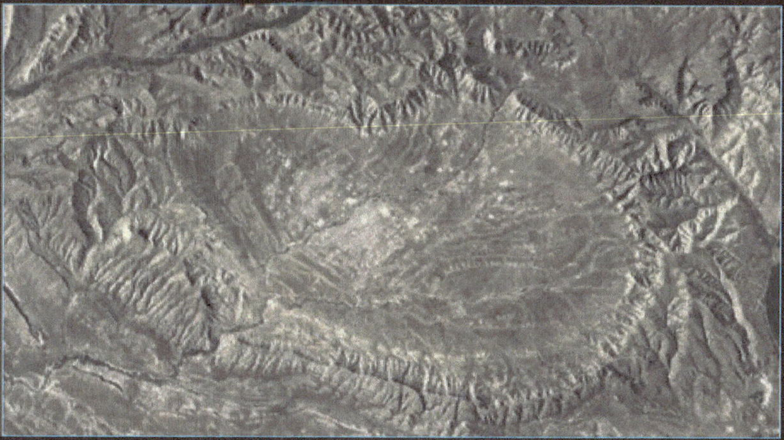

Exploring the vastness of Wyoming's geomorphology bridged the discovery trail to the Winds outcrops. By understanding the biomechanics, physics, and rocket species engineering features, the story began to take shape.

Dinwoody Lakes, Fremont Co., WY.

Fall 2020, Northern Wind River Reservation. The chain of 3 Dinwoody lakes was one of the most ominous alien adventures that I experienced over the last few years. First off, the lakes are still discolored from the Dinwoody Glacier; they are not known for trout fishing, nor are there any noticeable insect hatches or visible aquatic life. Also, I haven't seen the large wildlife like I usually do in other Wind River watersheds. The lakes are remnant, staged rocket mold filling with glacial ice. It's slowly clearing but still showing a cloudy, alien chemical life, outer space soup. There's no other lake that I have seen like this one in the Winds.

The first evening, I parked at the edge of a lake to gaze at the amazing petroglyphs fenced off. Then, I noticed a pod of trout rising, feeding on something just below the surface. It became a feeding frenzy, but I couldn't see any insects flying or floating as I tried a couple of fruitless casts. I had to go back the next day. I was so curious, and the light was too poor for best rock pics on the first day.

When I arrived the next day, at the shoreline, there were three fresh dead trout not there the night before. I couldn't understand it or see anything like it. At this point, I had to walk around the lake corner, where there were multi-colored outcrops and small dark caves with fresh entry tracks. All alone on the eerie lake, I felt that I was being watched as an intruder, but I still had to take better pics of the alien artwork.

Driving home the next day, I had to call Bob, my Shoshone, UW buddy, to tell him about the spooky episode. He then shared this story about Dinwoody, "That's the reason why my grandma told me never to go there because of the Nimerigar, 'Little People.'

Over these two days, I saw things that instilled in me the fear the natives felt when they saw the creature Nimerigar.

Finally, I think the small stocker trout fed on the hatching, swimming aliens in the lake and died overnight, not able to digest them.

Driving home the next day, I had to call Bob, my Shoshone, UW buddy, to tell him about the spooky episode. He then shared this story about Dinwoody, "That's the reason why my grandma told me never to go there because of the Nimerigar, 'Little People.'

Dinwoody, Wyo.
petroglyph figure
50cm

EIN embryo fossil
1cm

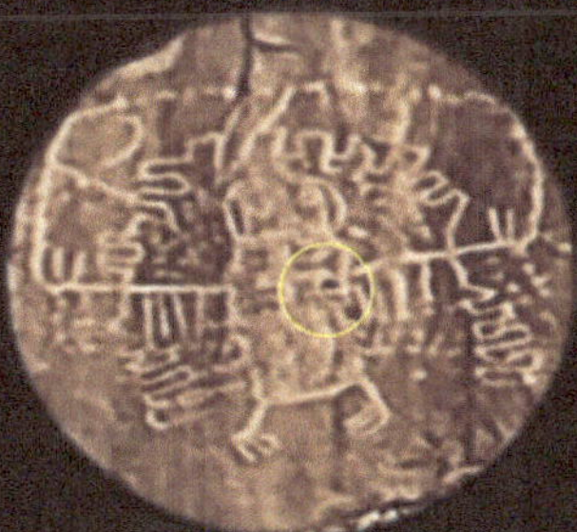

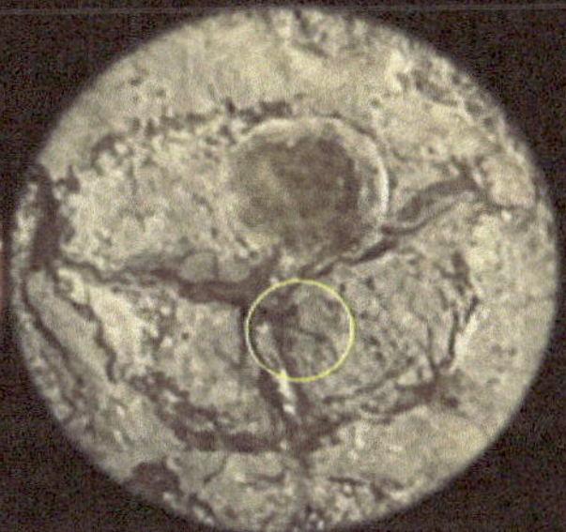

Tubes, body parts around heads, bodies, extended arms. Circled is the H20 fuel cell/battery - this vital body part has been found (2mm – 8cm) in EIN skeletons hundreds of miles away in CO.

THE CHASE

The discovery of active burrows at the Wyoming crater sites ignited a thrilling chapter in my exploration fever, beckoning adventurers to Colorado's plains in pursuit of elusive subterranean species. Such a quest was driven by the real possibility of unearthing living entities that defy current biological understanding, perhaps even extraterrestrial in origin. The narrative is further enriched by historical accounts, such as the peculiar find within the Pedro Mountains of Wyoming—a diminutive, mummified figure that sparked debates and captured imaginations. This 'Little Man' mummy, as some have called it, was discovered in the mid-20th century and has since been a subject of controversy and fascination. It's a tale that intertwines the academic rigor of anthropology with the allure of the unknown, challenging the most skeptical minds with tangible evidence from the fossil record.

The lore of North American Indigenous peoples adds layers of mystique to these investigations, especially given that friends and family appreciate the regional gravity of the evidence. Stories passed down through generations speak of small, formidable creatures associated with water, evoking images of a world where the lines between the natural and the supernatural blur. These narratives, rich in cultural significance, often depict interactions with beings that embody the primal forces of nature—entities that are as fearsome as they are revered. Such tales are not mere fiction; they are integral to the worldview of many Indigenous communities, reflecting a deep connection with the land and its many inhabitants, seen and unseen.

In the modern pursuit of scientific discovery, these ancient stories take on new relevance. They remind us that our understanding of life is constantly evolving, shaped by both empirical evidence and the enduring wisdom of traditional knowledge. As explorers and scientists dive deep into the earth's hidden recesses, they do so with a sense of wonder that has fueled humanity's quest for knowledge since time immemorial. Whether seeking unknown species in the shadowy backroads of Colorado or recreating the origins of a mysterious mummy in Wyoming, the journey is as much about the search itself as it is about the potential discoveries that await. It's a testament to the human spirit's insatiable curiosity and our never-ending dialogue with the mysteries of our world.

The Rainbow Rocket was spotted on land east of Denver on the first evening back in CO. I went to examine it, but the rocket was on private land. I've since tried to re-find the 70' rocket even after I staked in a map file. This was the first object that had details of a reactive coloration and flight structure.

The public park adjacent to the plane, however, was the first in a series of astonishing and mind-bending trips to the SE Colorado plains.

Colorado SE plains revealed more defined features. Walking, starting east of Denver at the circle object area.

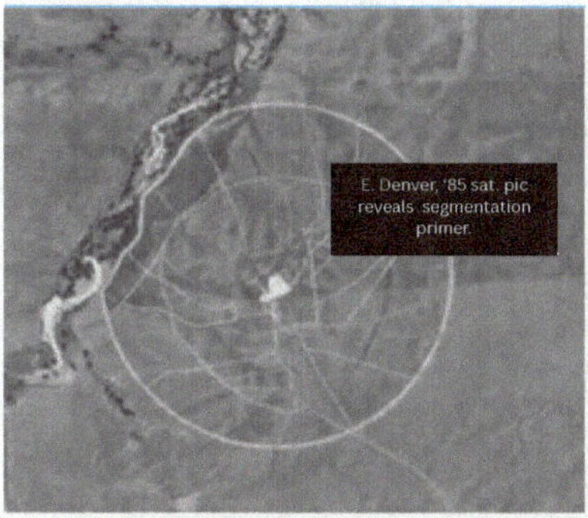

This public site was the first stop in Colorado and was a treasure trove of evidence that revealed cement-looking geometric moulded slabs and a carbon-based metal water tank that I tested in the lab. This site had deep ground holes and a non-working wind-mill water well pump.

On the water well, I found what I first thought was a dead mouse. Without odor, I knew the 3 in. skeleton, bones were not of any animal species.

Two weeks later, same site, I found 2 more skeletons; one the same size, the other 1" long.

During the same trip, smack in the middle of the terraform water tank was this rubbery, earth-camo piece of material. It was probably to protect and carry creatures from deep pressure fissures, migrating up through old water well bores. It also tested organic Carbon.

I saw more evidence of the species taking over other abandoned water well bores in CO. One had an electric fence with no reason other than using the tank and bore to focus an energy source.

Rocket parts, next to a beautifully clear blue-green water tank (next page background) was like never water that I've seen anywhere, let alone on the dry plains of CO where surface water is a rare source. This small pond only developed recently and is perhaps is an alien bath, breeding water with deep c rcuitry.

Fresh, active EIN burrows, 1" to 8" dia., were found freshly dug in hard roads along with a sample of black translucent eggs like the green WYO ones. The 4" wide claw marks a 5-finger hand. After I found the obvious signs, I won't stop looking at the markers to this day, like spreading mortar cement in the impactor dam construction.

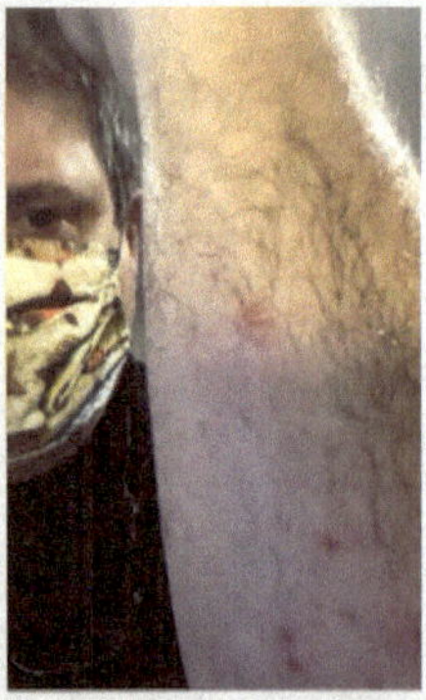

The carcass clumps were devoured and regurgitated by one larger creature, and then I dissembled and reassembled them, building the anatomy with the copious parts. This day, I felt, however, the acute pain of the alien darts in my forearm.

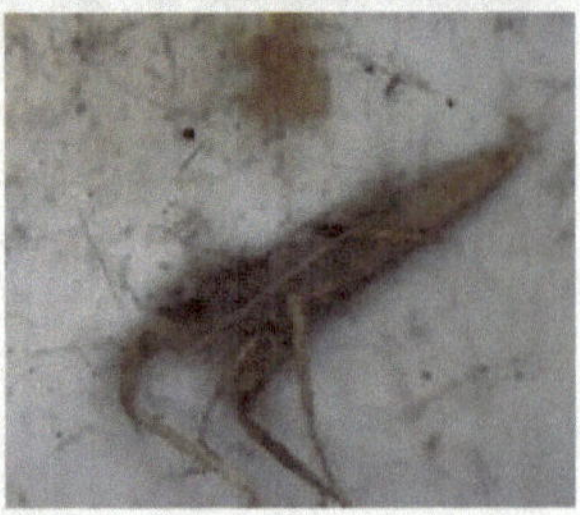

The poisoned-tipped weapons apparently landed on my armchair while working. This episode was so painful and primitively horrifying that I was prepared to take all the evidence and bury it all back in the park!

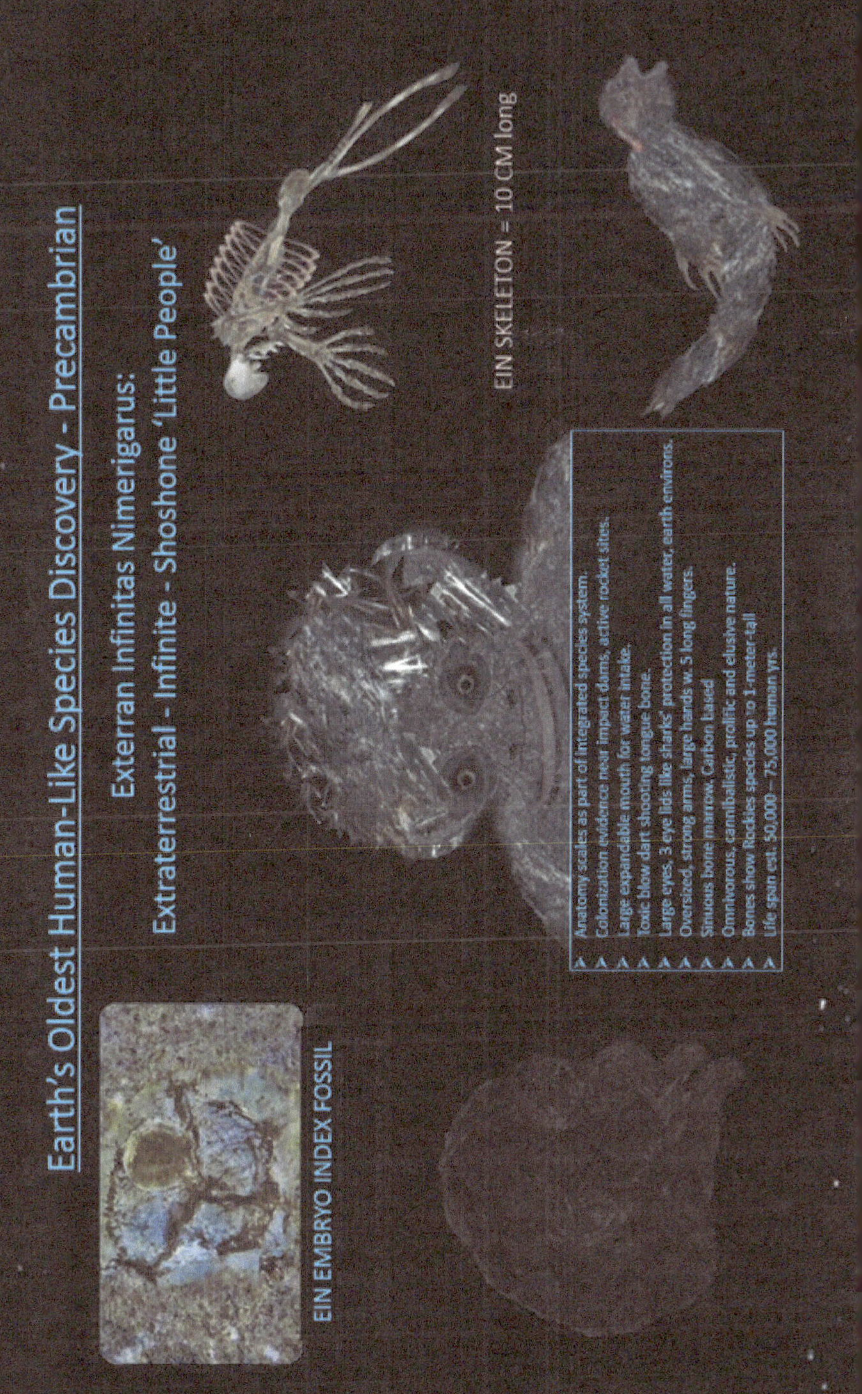
Earth's Oldest Human-Like Species Discovery - Precambrian
Exterran Infinitas Nimerigarus:
Extraterrestrial - Infinite - Shoshone 'Little People'
EIN EMBRYO INDEX FOSSIL
EIN SKELETON = 10 CM long
Anatomy scales as part of integrated species system.
Colonization evidence near impact dams, active rocket sites.
Large expandable mouth for water intake.
Toxic blow dart shooting tongue bone.
Large eyes, 3 eye lids like sharks' protection in all water, earth environs.
Oversized, strong arms, large hands w. 5 long fingers.
Sinuous bone marrow. Carbon based
Omnivorous, cannibalistic, prolific and elusive nature.
Bones show Rockies species up to 1-meter-tall
Life span est. 50,000 – 75,000 human yrs.

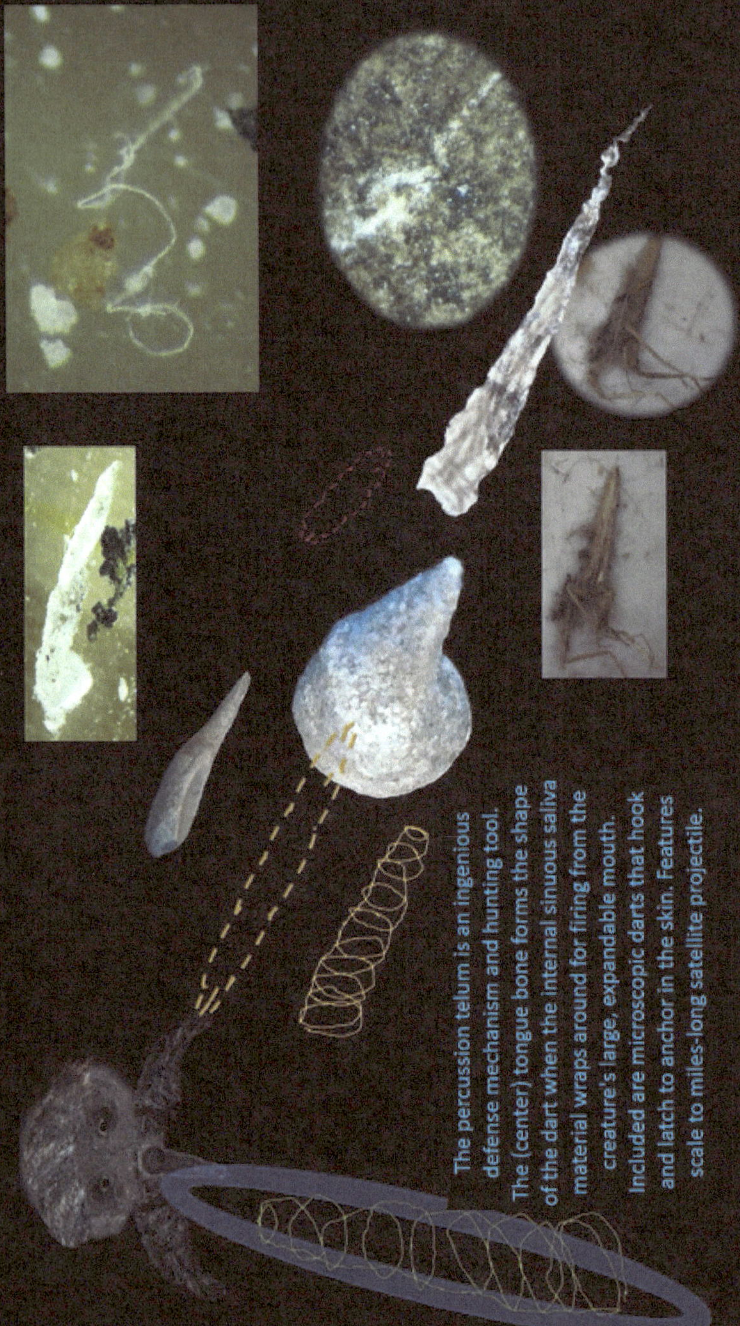

The percussion telum is an ingenious defense mechanism and hunting tool. The (center) tongue bone forms the shape of the dart when the internal sinuous saliva material wraps around for firing from the creature's large, expandable mouth. Included are microscopic darts that hook and latch to anchor in the skin. Features scale to miles-long satellite projectile.

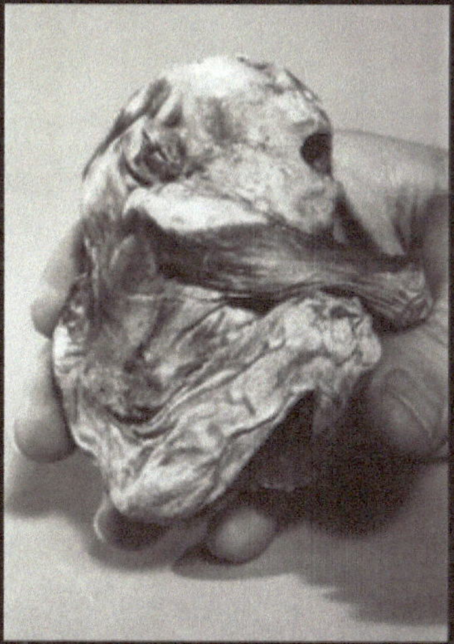

WYO specimens: La Chiquita mummy (l), Pedro Mtn. Mine, Little Man Mummy* X-ray, helped me solve the anatomy, and not only as an aggressive digger but after I made the cut out, it made total sense it was a swimming, slithering creature with a body cavitation. It has big eyes and three (3) eyelids like sharks with a controlled gaping mouth for water propulsion and dart blow gun percussion energy. Sideways, the creature is flat like a dragonfly nymph that uses body propulsion.*
Museum of Nat History, Harvard Univ. of Wyoming photo credits.

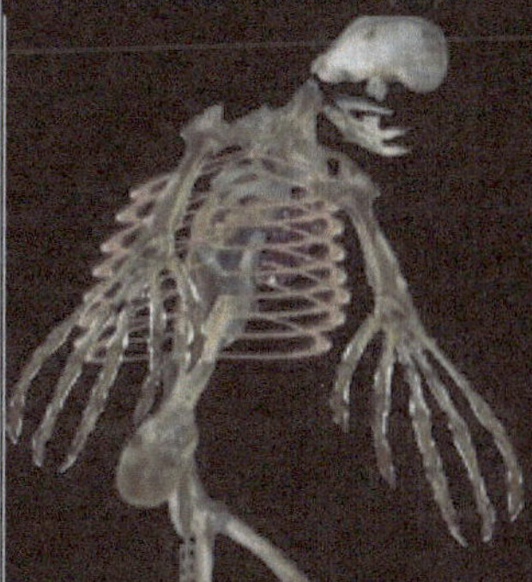

It matched the fossils to petroglyphs and the Wyo. Mummy x-rays, to the CO skeletons, were a definitive piece of the puzzle, solving their purpose based on its colonization, habitat, anatomy, and digging, programmed to manipulate the surface and rebuild the master rocket species forever.

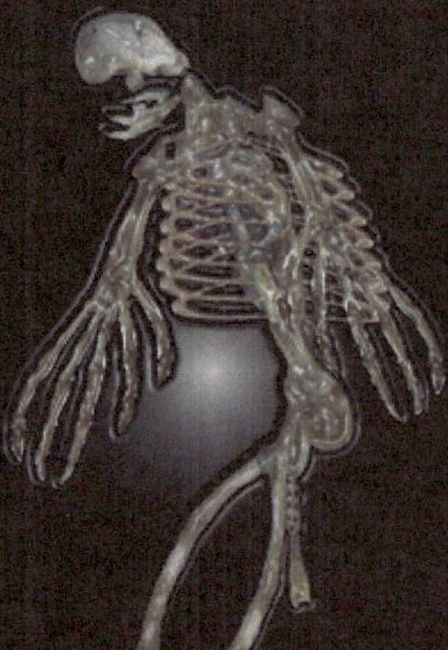

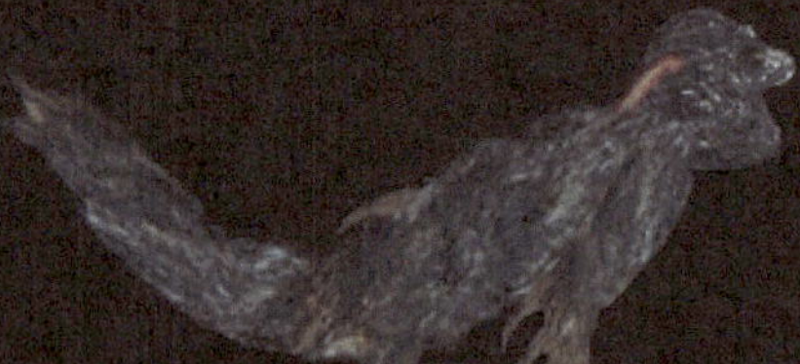

The SEM bone results. The head, skin, jaws and pelt were cropped to make early figures.

Element	Mass%	Atom%	Sigma	Net	K ratio	Line
C	15.89	26.41	0.03	19266	0.0154549	K
N	4.44	6.33	0.09	2137	0.0135901	K
O	35.54	44.34	0.09	67238	0.1586522	K
Na	0.64	0.56	0.03	3074	0.0065336	K
Al	1.44	1.06	0.04	7489	0.0179158	K
Si	2.59	1.84	0.05	12890	0.0353457	K
P	11.16	7.19	0.09	43598	0.1243185	K
Ca	21.79	10.85	0.16	41870	0.3245976	K
Zr	6.52	1.43	0.23	16439	0.0645524	L
Total	100.00	100.00				

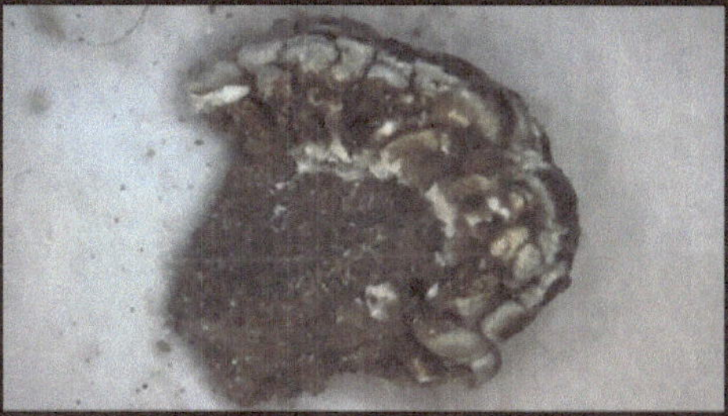

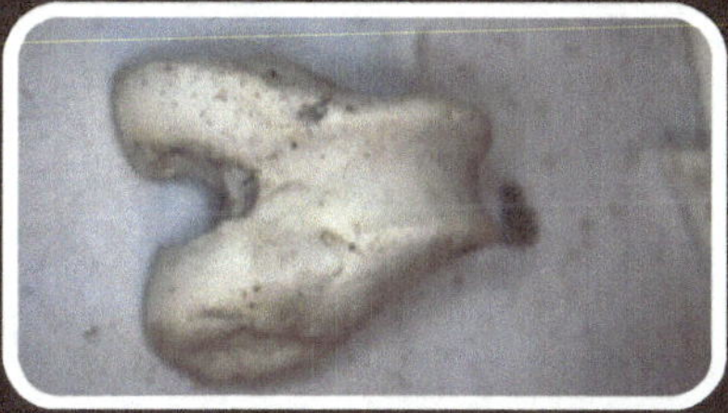

Small skull shows lung grove

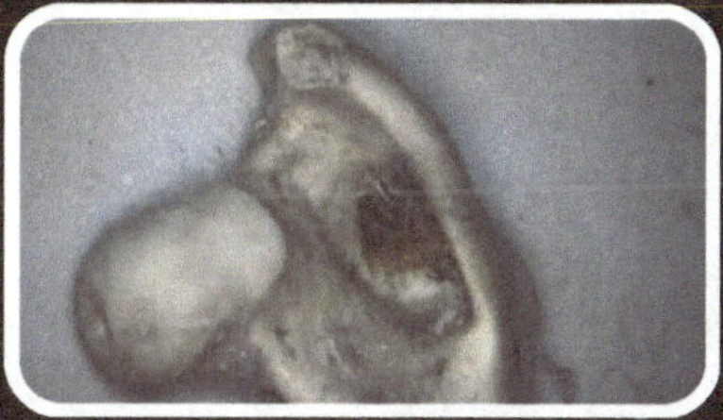

Water battery chamber

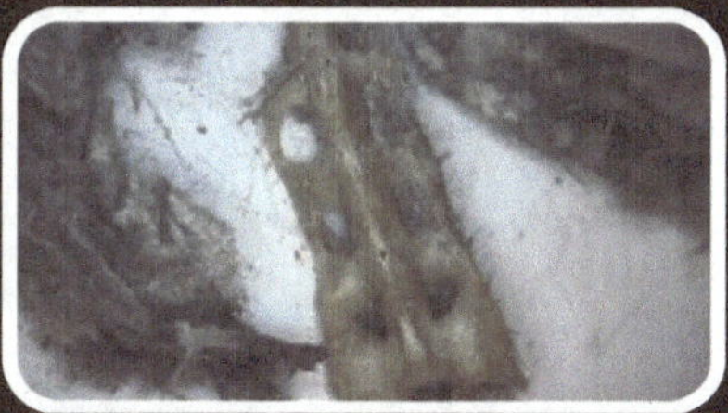

Articulated tail bone

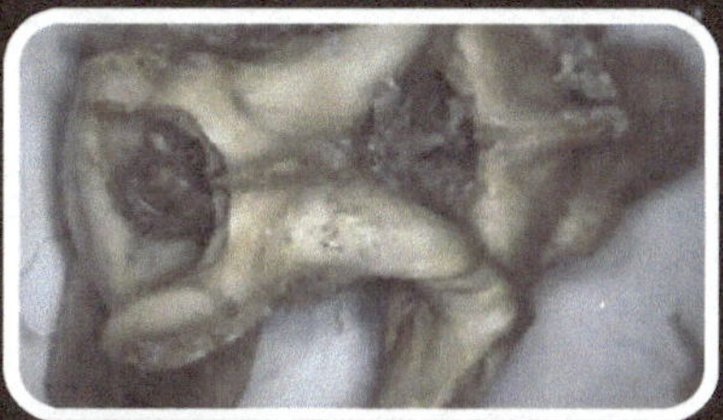

Torso

Curved body-blade

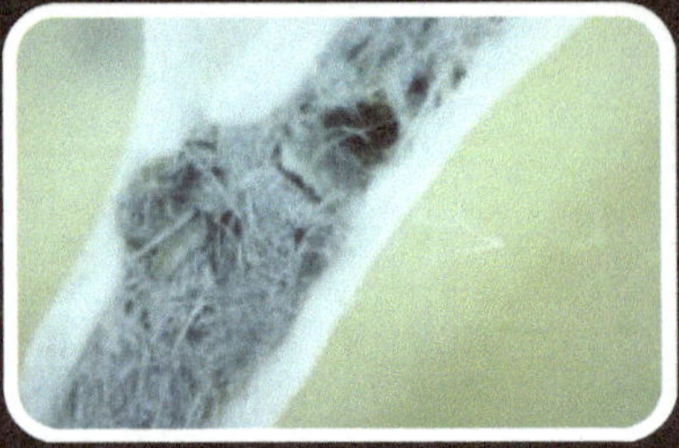

Leg bone shows sinuous marrow material

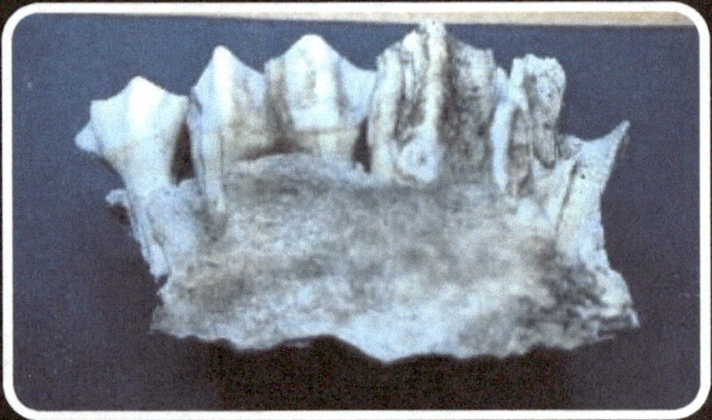

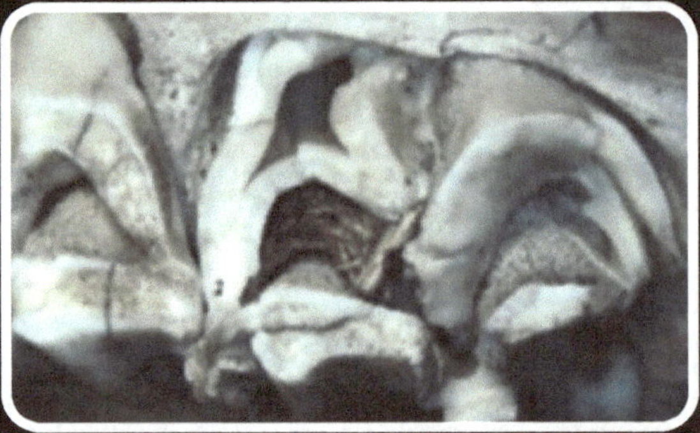

Teeth cavities storing 3 different mixing materials.

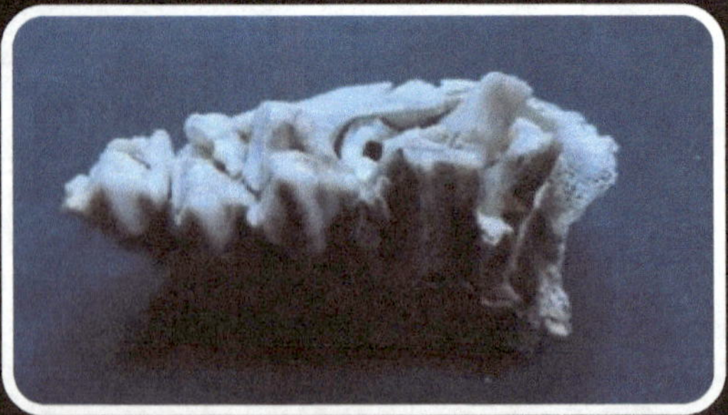

The EIN jawbone is also a rocket

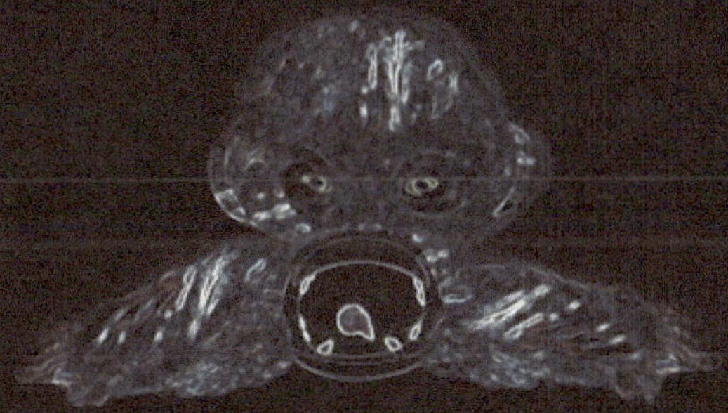

From the first image without skeletons to draw(!), to the one being threatened, it was a most fascinating aspect studying the anatomy of the elusive species. More bones haven't been found by others because they dissolve, returning to Earth and regenerating. Due to the same species, no Big Foot, Swamp Creature, or other skeletons have been discovered.

This species is heavily concentrated in the Western US.

Jones Pass, CO., 2021, I spotted this 3 ft. EIN tunnel on other side of a stream. I knew it was a body-blade drilling team effort into granite. This home cave is at base of the Rocket Mountain Factory discovery.

The cave of the Little Man, Wyoming 1932, a 15" mummy was found in a similar cave in dimensions and precision.

The earliest EIN species may have been more water bound, metallic and robotic but may have changed color in higher oxygen levels, temps and with the evolving marine and land food sources. Fossils, literature depict the creature with an aqua lung, but the feature was difficult to render properly.

INVASION

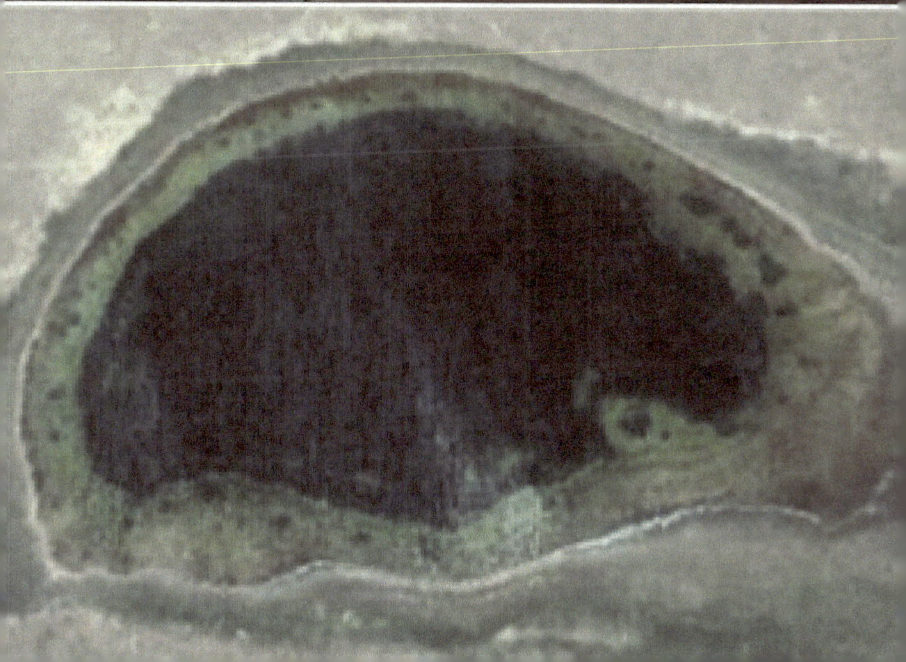

The chapter evolved my thinking into the flight, landing mechanics, and astro-geological exobiological aspects of the species invasion structure.

Since the bombs hit the surface repeatedly and locally, I had to know the species' flightpaths, landing, and sprawling patterns. With enough data assembled in the Western US, it was then traced all over the world.

The surface patterns should become recognizable to many people, especially in the Western US. More will study the routine impact of dam lakes on their travels, in their neighborhood parks, golf courses, and perhaps on their own land.

To date, bombardment events have only been theories for any of the 6 greatest mass extinctions because of the changes in conditions recorded. But what hasn't been recorded until now are the extreme event resources that caused these severe changes.

During this time of research, before I found the Invasion symbol, I still didn't know what the bombs were about or shaped like, which were causing such vivid surface deformation on the Plains. I used the projectile embedded in the Aquamarine alien gemstone as it was the only solid lead at the time to understand the landing and penetration characteristics.

Countless alien bombs hit the Western US and the world, forming impact or lakes. They are encased with a fortified outer metallic shell that penetrates the surface violently with a spring-loaded landing foot. It then spreads to create a water control system where the matrix habitat mineralizes out of water, and the subspecies workers hatch. The stealth bomber design had already been utilized billions of years ago!

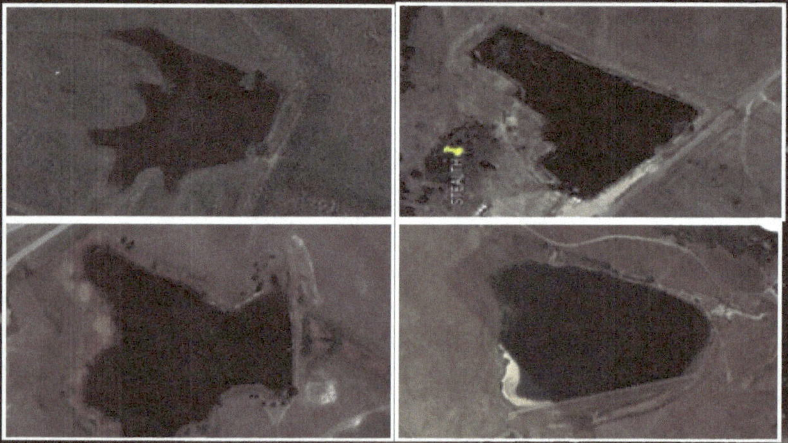

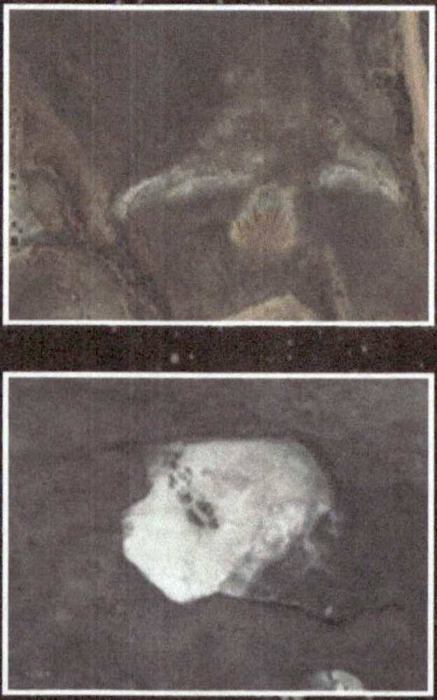

Impacts exhibit the species outer hardened shell casings.

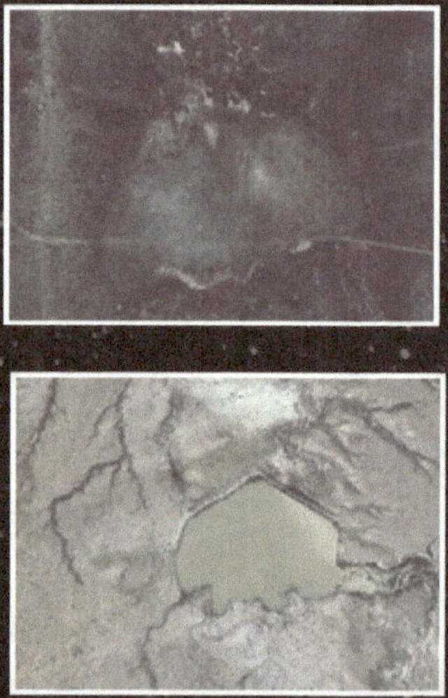

Impactors are encased with a fortified outer metallic shell that hits surface with spring loaded landing foot that buries deep in surface, then spreads to create water control systems where the matrix habitat mineralizes alien with water.

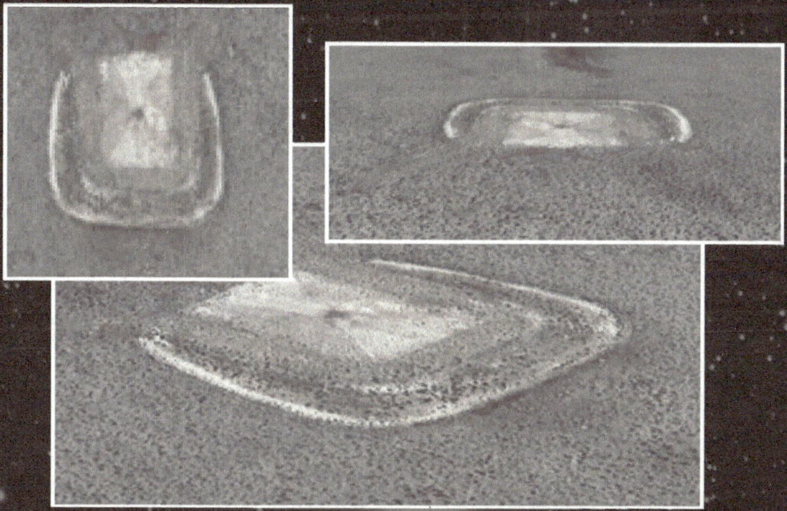

Cross section of a WY landing exhibits the mechanized spring loaded telum, earth dart release and penetration mechanics burying alien anchored as deep as possible to live in earth. Species release long barbed tail expanding, unfolding seeking H20. Site estimate landing @ 10-15kya ago.

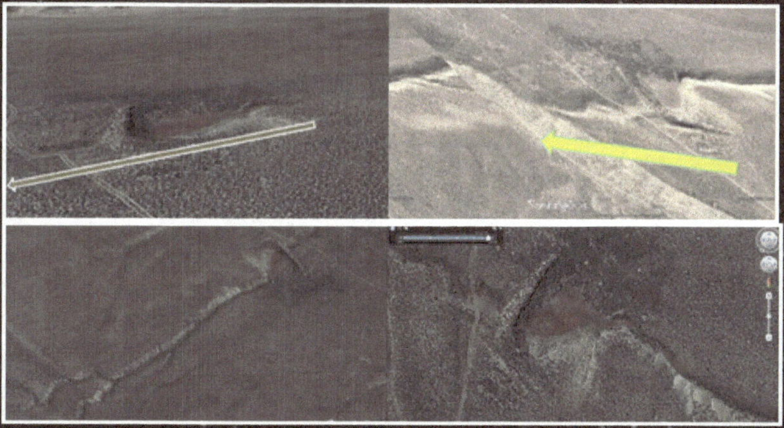

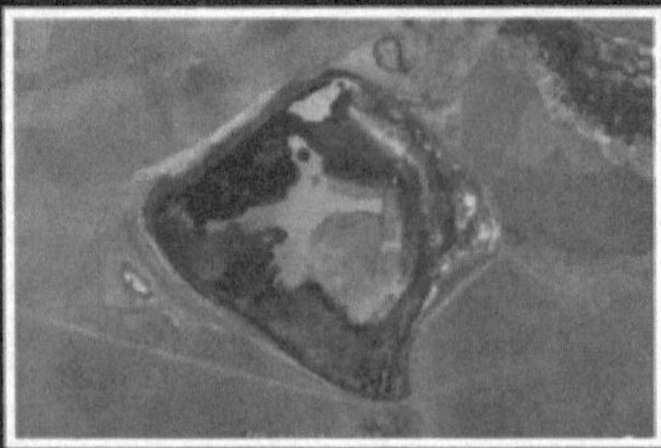
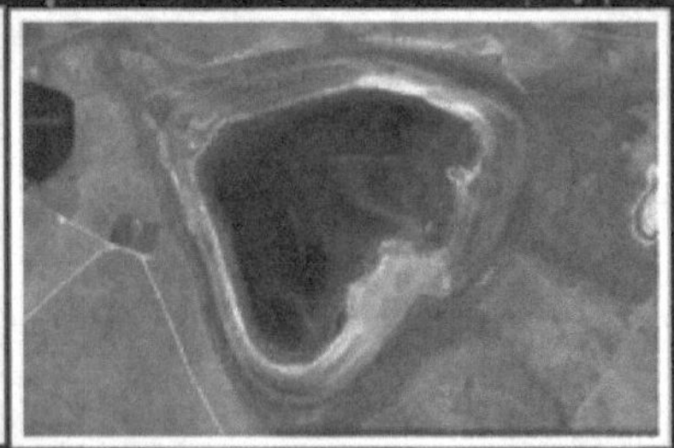

It was relentless bombardment episode with signs everywhere. Another ongoing challenge is to stop taking pictures and managing the images to present appropriately for the audience.

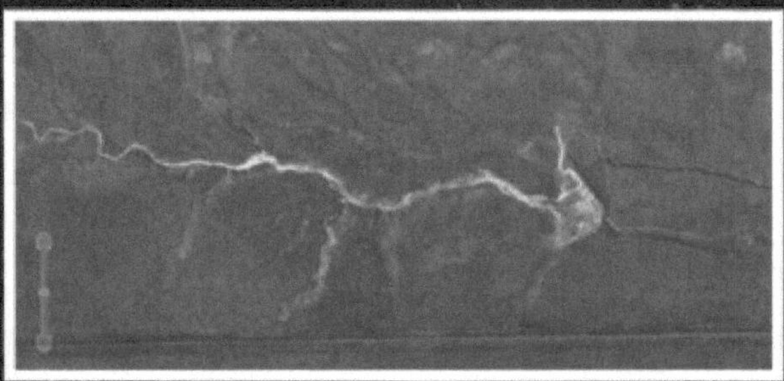
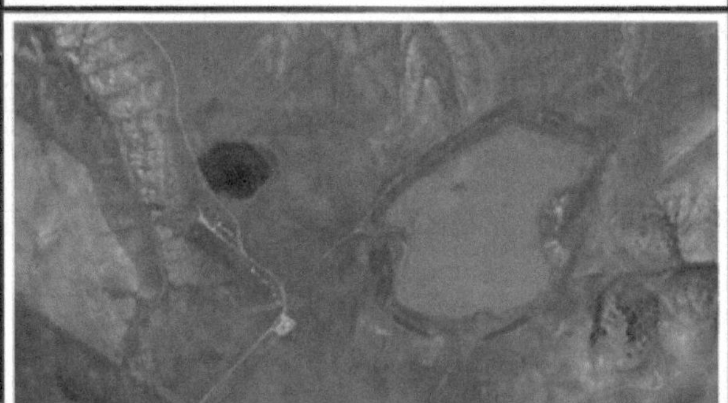
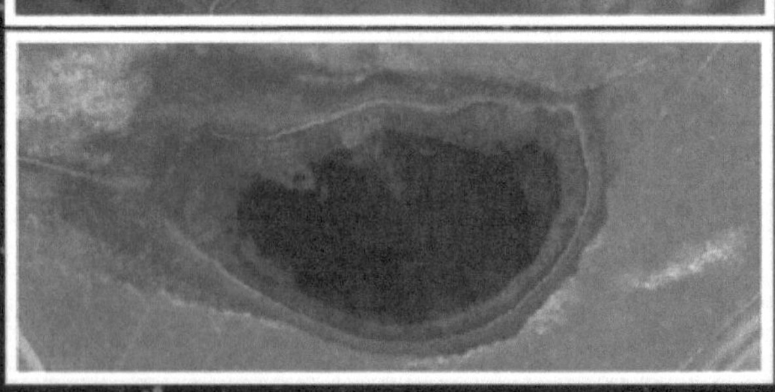

South America - Alien bomb smashes into soft earth.

WY site exhibits landing direction, penetration apparatus, early dam structure process.

<u>Exterran Infinitas Erucae Animantis</u> (EIEA); Extraterrestrial Infinite Organic Rocket.

N. WY - The discovery of this 30 mi. wide impact species, its internal characteristics became routinely recognized in the field and on satellite to identify the species common impactor traits. At first, I was certain that had just found the largest fossil in world, but they got bigger. This is Top 10 photo.

Out of this world landing exposes massive rocket specimen with eyeballs and body that exhibits the great sting ray design. By studying the eyes spirals, contours I could find the alien eyes all over the globe on the surface or part of the city landscapes.

Southern Chile specimen was another discovery breakthrough, seeing how the strong metallic arms stretch and segment, the unfolding occurring over time from super compaction. Other specimens in the area look like huge alluvial fans.

Landing high in the Bolivian desert, this alien bomb site exhibits a thick, extremely hard outer shell layering, splintering and expanding.

A blue water sac, electronic circuitry and surface alloying (grid matrix) are visible.

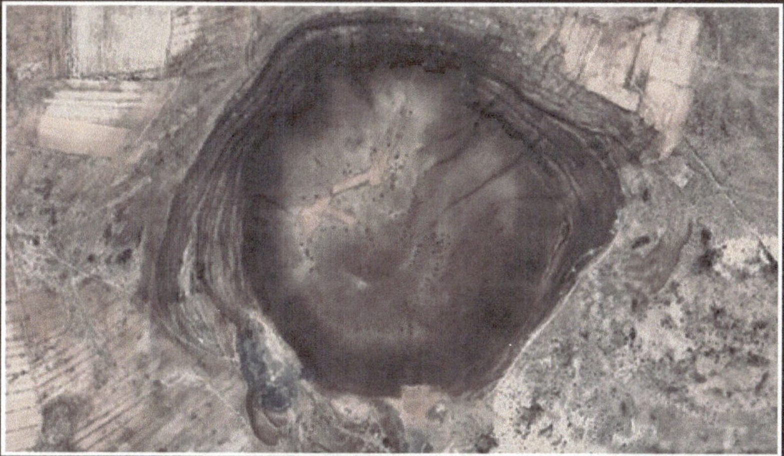

Delta UT - 2 sq.mi. impact site exhibits alien impact species with characteristics of the force and reaction of its landing mechanics. This causes shattering outer metallic layers, causing domino effect for expansion outward to dam water and live forever.

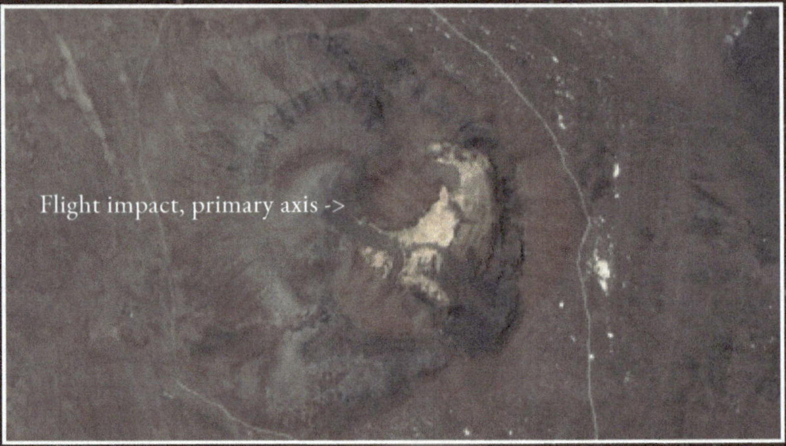

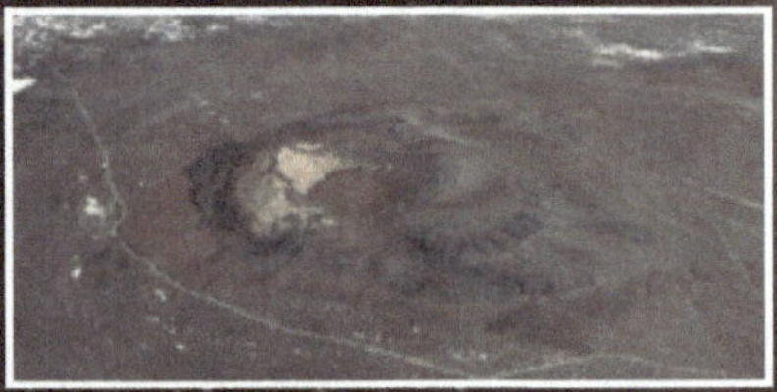

The Lace Curtain, Delta, UT

These distinct, unfolding outer layer features were at first only anomalous geological processes but help characterize the sprawling nature of the species. Mores book are necessary to show the vast amount of data.

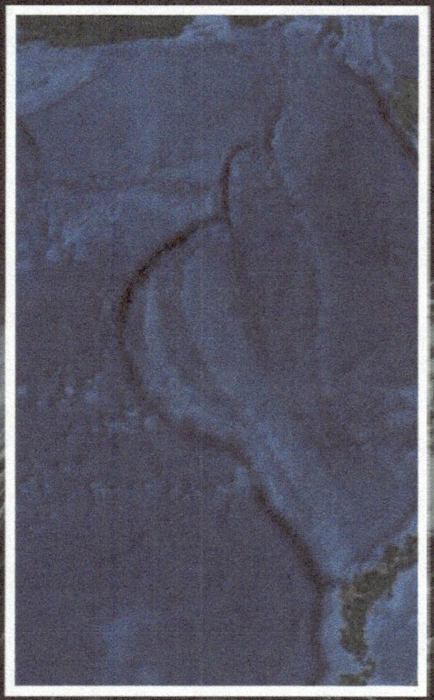
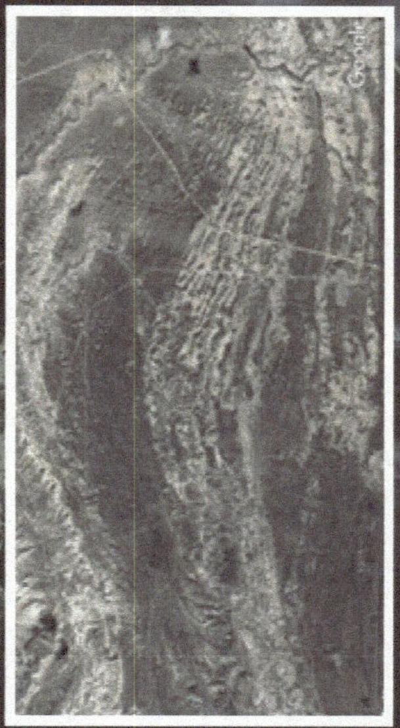

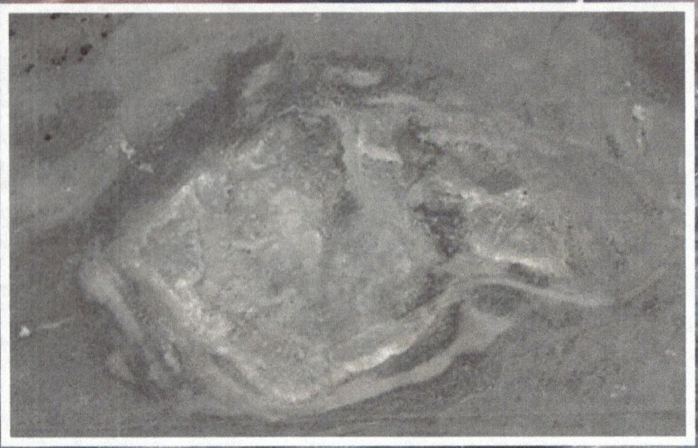

WYO EIEA species show similar bony mouth segmentation, opening to the giant grouper. These features resemble past and present marine and land animals.

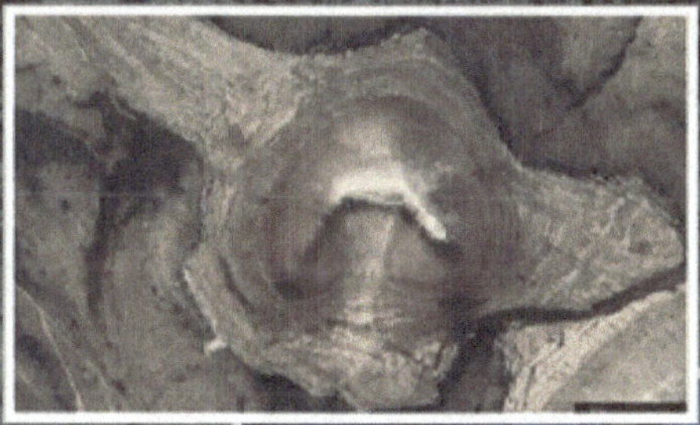

The distinct speciation and the resource patterns is explained with Geoscience but not theory.

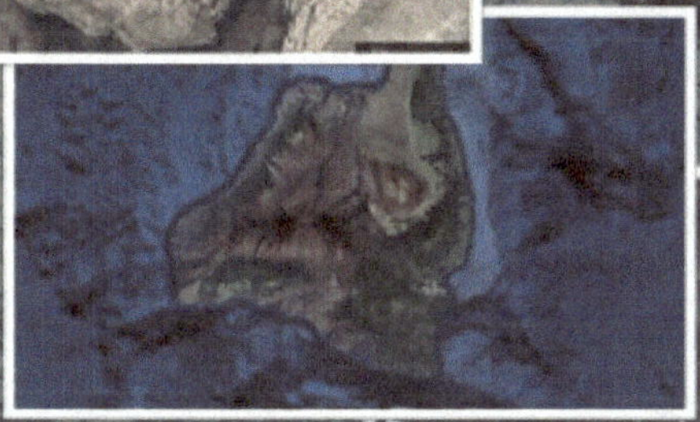

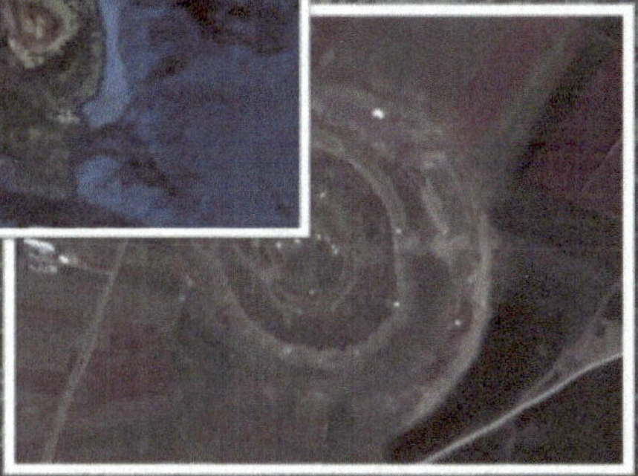

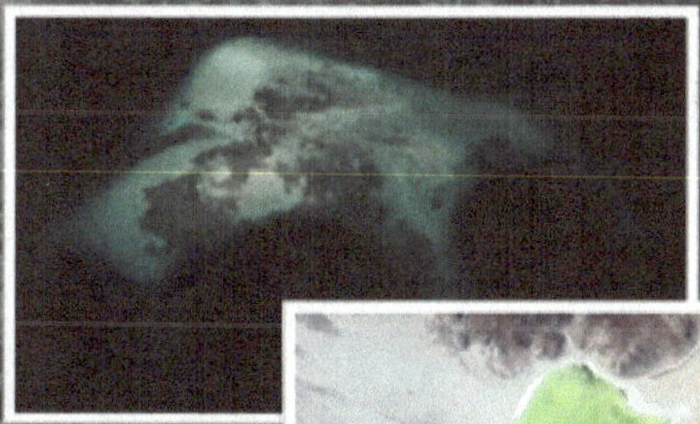

Evidence is found everywhere on Google Earth, in all environments around the world,

and it's all about water. Since ambient water resources determine the dormancy and growth cycles on earth, the work should prove to connect more dots.

Atolls, and islands across the oceans are not all volcanically nor microbial deposits created.
The land mass is the outer space invasion repeatably seen as geological remnants.

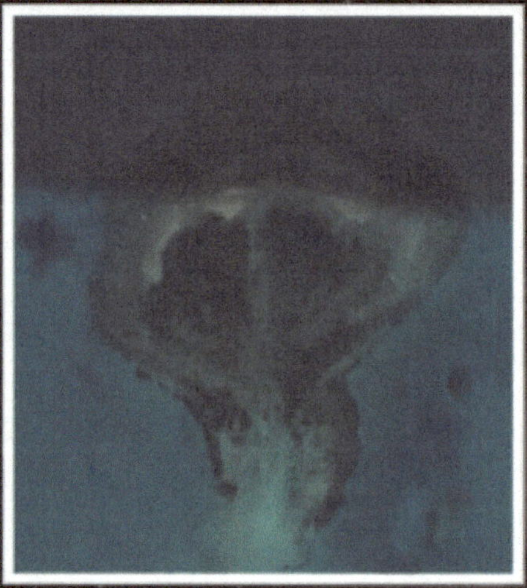

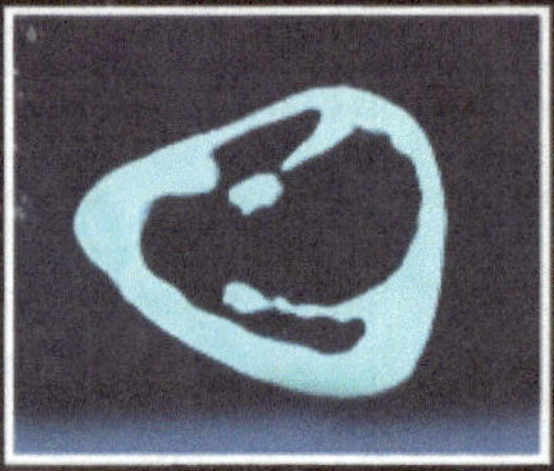

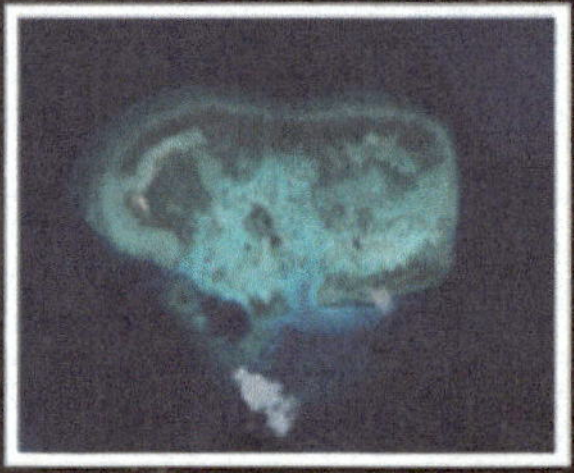

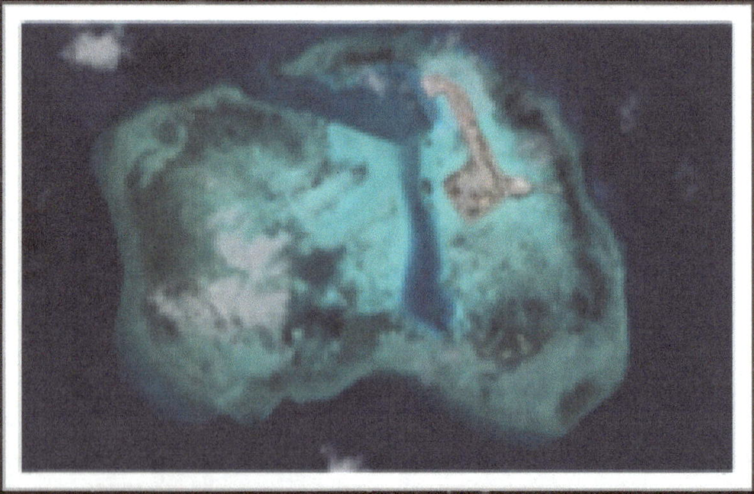

The last Mass Extinction; Holocene Epoch 10,000 yrs. Ago.

Constructive, metallic attributes are ubiquitously seen the older species.

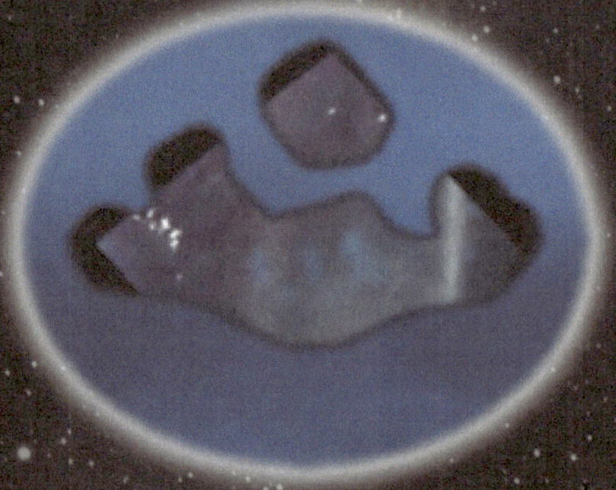

Life existed on Mars and is illustrated by the same speciation, impaction found on Earth.

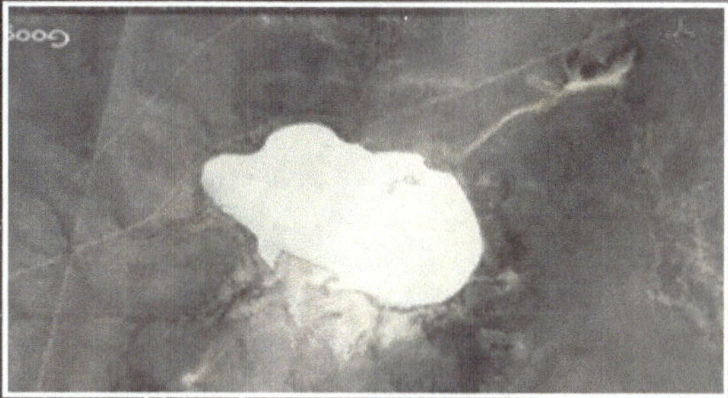

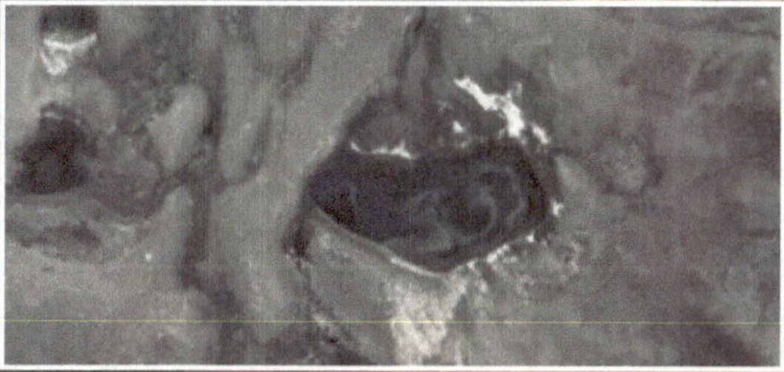

NASA Perservance Rover, Sept-2022

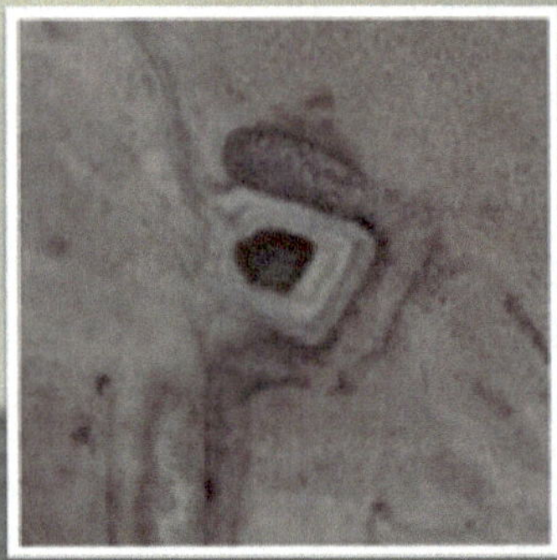

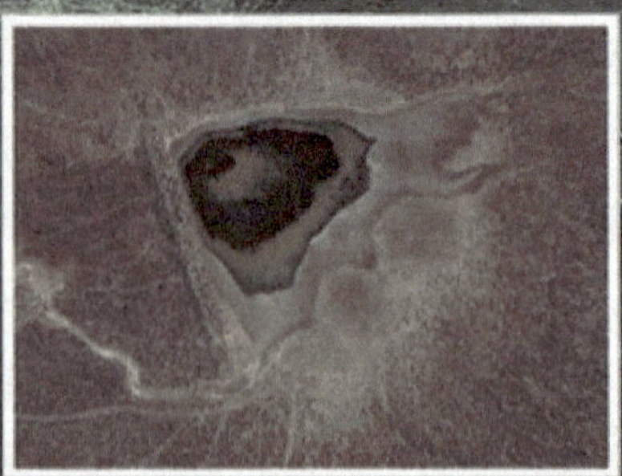

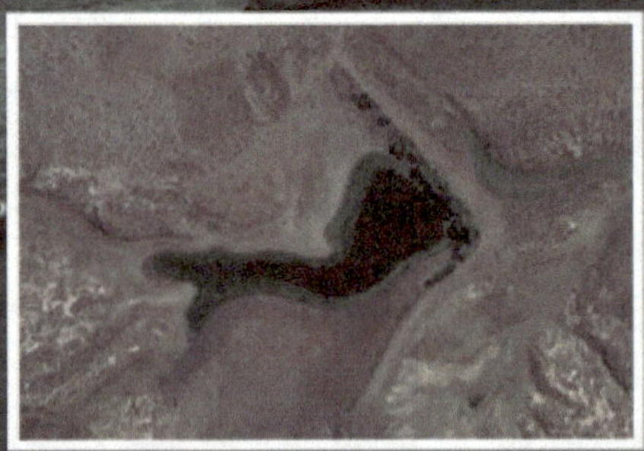

Alien impact dam features are ever unfolding, rocket re-construction sites seen @ 18,000' in Himalayas and in the Rockies.

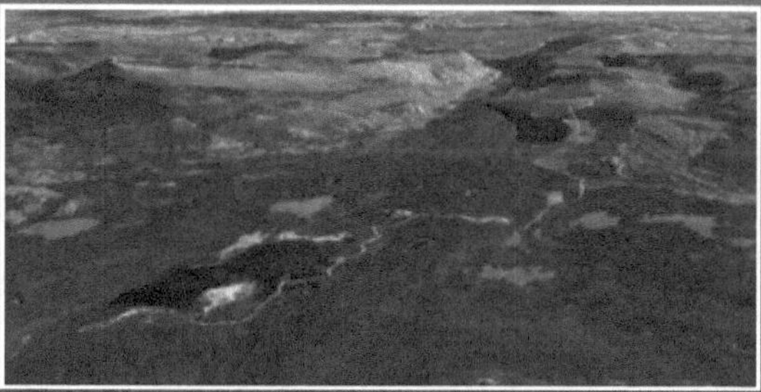

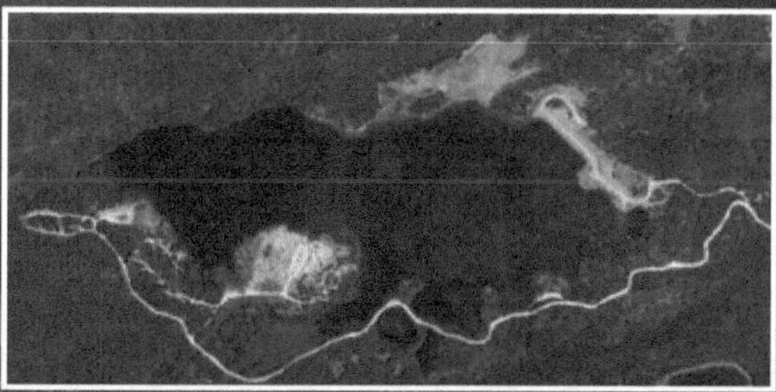

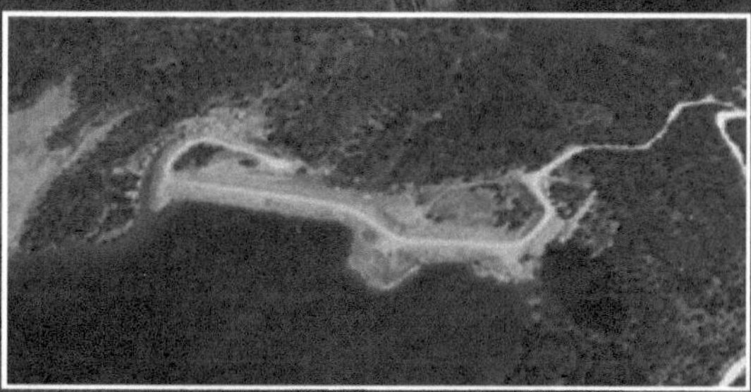

After performing a Frye Lake, Worthen Reservoir cost/logistical analysis, it was determined that the dams are 100% non – human built given the rock tonnage loads, money and with limited construction time constraints atop a 10-mile winding mountain road to the twin 9000' foot locations. They are simply not feasible and were not built for irrigation storage, flood control nor have economical use nor have any traditional mitigation, human engineering tenets.

Twin impact lake features in WY expose the high sophistication and stellar circulation of outlet patterns.

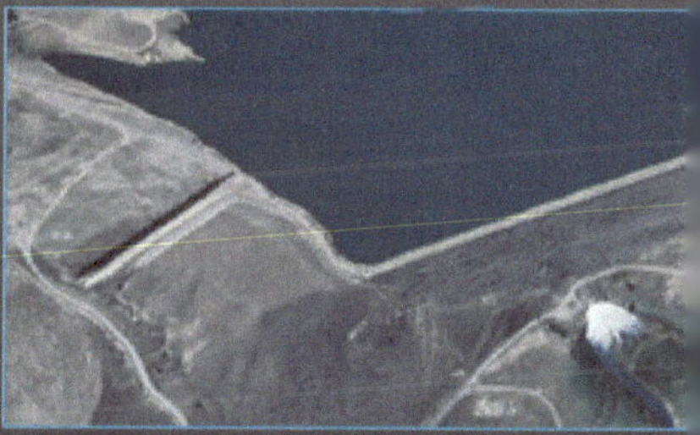

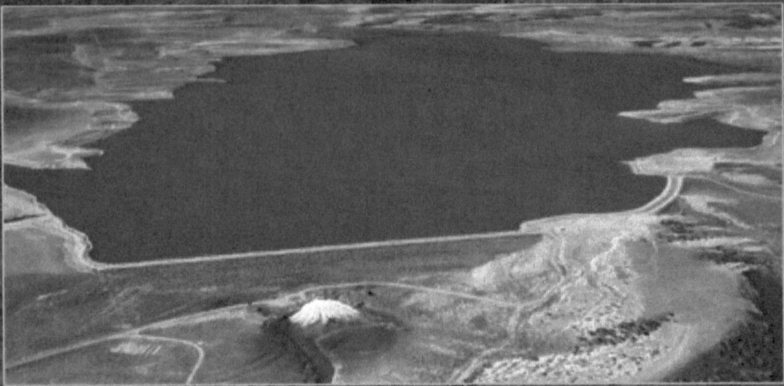

A deeper, more sophisticated layering structure seems to be the most effective way to estimate the age of the invasion of the formations. These seem late Precambrian, Early Cambrian. I've checked a few of the larger reservoirs in WY in the National Archives for dam permits. However, it was a waste of time and not practical because these reservoirs were not man-made after all..

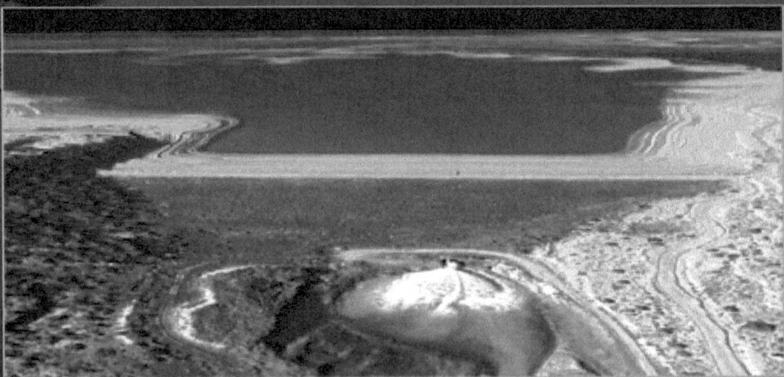

Some of the largest non man-made WY reservoirs exhibit an outflow manipulating outlet pipeline systems that feeds miles of the desolate lands across the WY plains. It's all about the water volume storage and chemical reactions to scale.

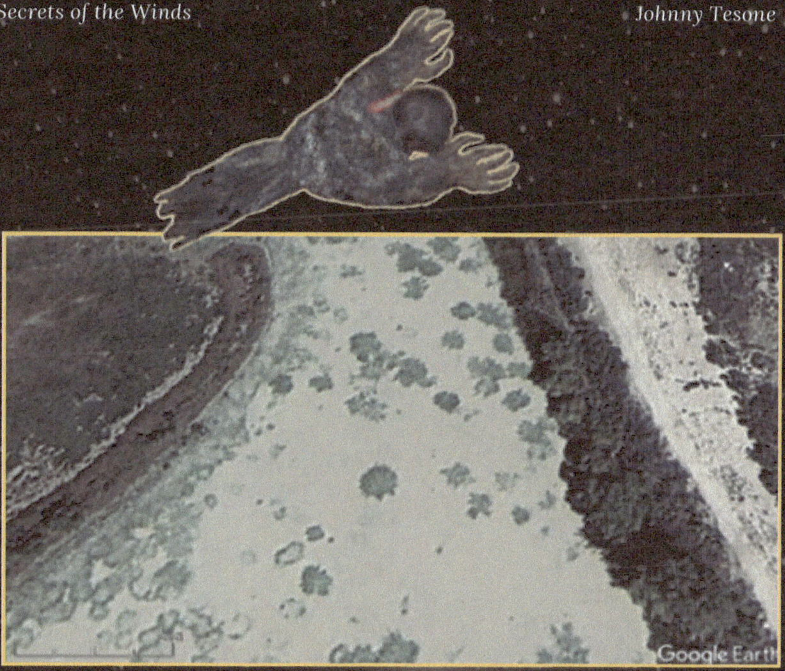

Impactor dams have associated past and present burrowing mounds worked and inhabited by the EIN creature.

Cobalt Crater Discovery

The massive crater exhibits a macro petri dish example of the superalloy process and acceleration properties, particularly substantial over the last 20 years.

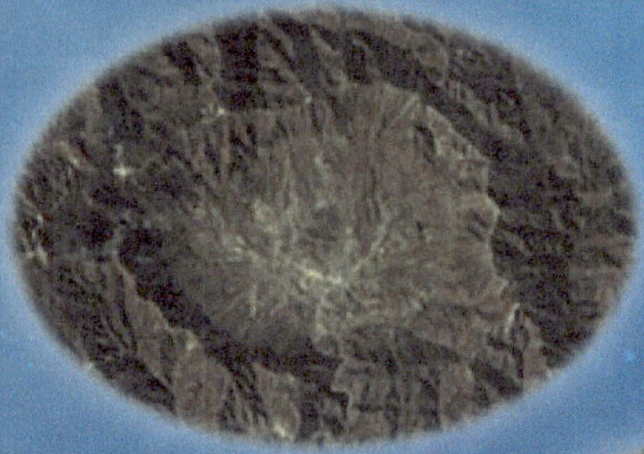

↑N, 1991

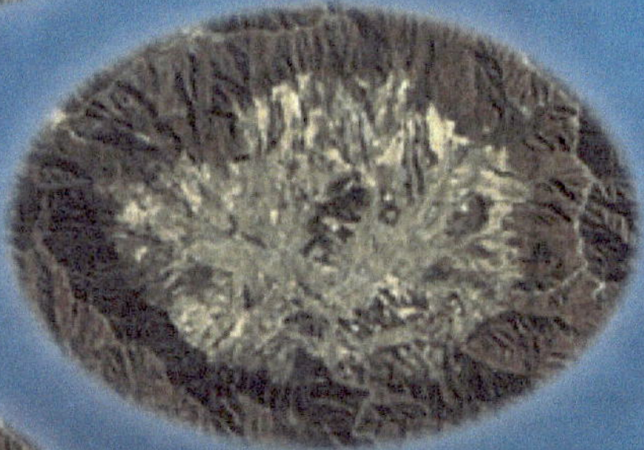

2005

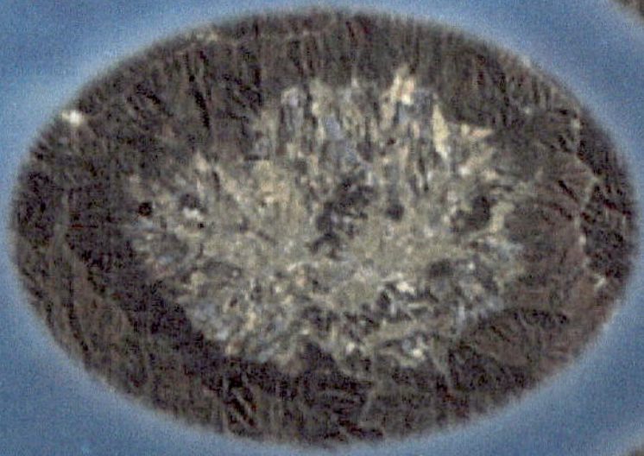

2012

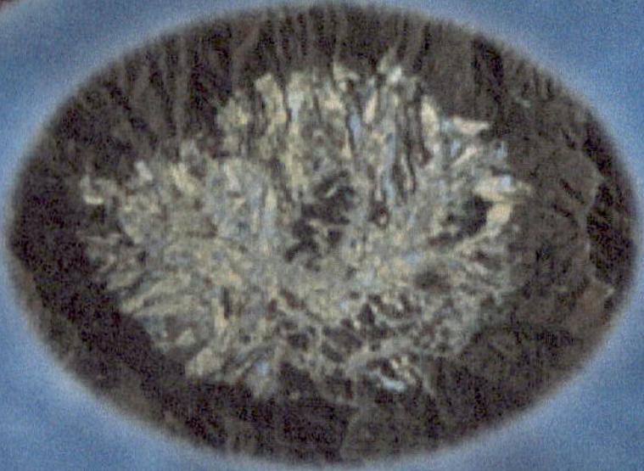

2024

Landing swarms are evident all over the world and have determined they are repeatedly launched from the rocket ships canons.

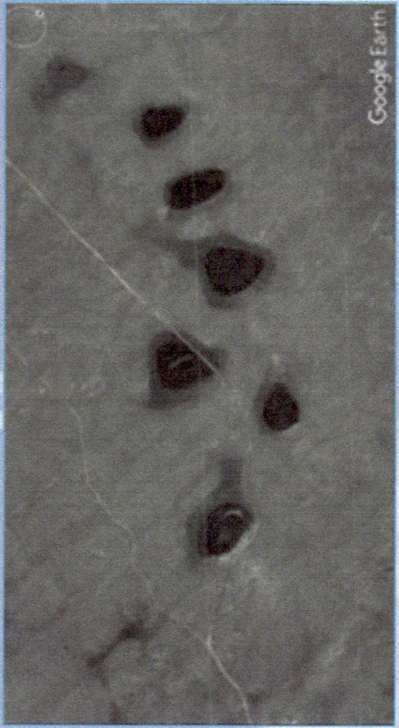

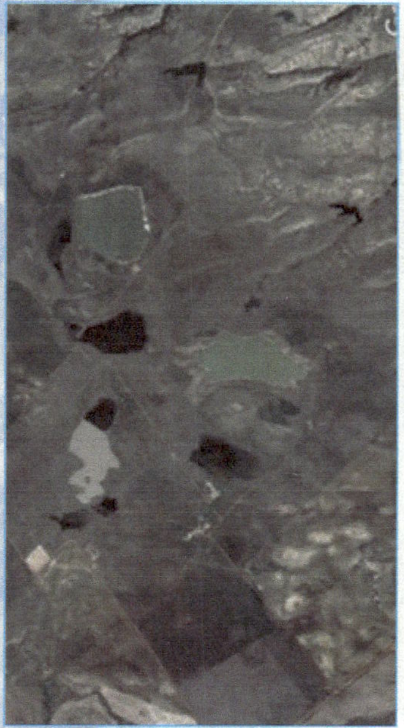

Jefferson Co., CO

Impact lake investigation of dam structure, and the rock mortar substantiated the previous 2 plus years of work on the Invasion chapter. A year earlier, in 2021, I also discovered active EIN mounds.

Driving along the front range of Colorado to Central Wyoming, the features are unmistakably abundant that even my friends and family believe in them now.

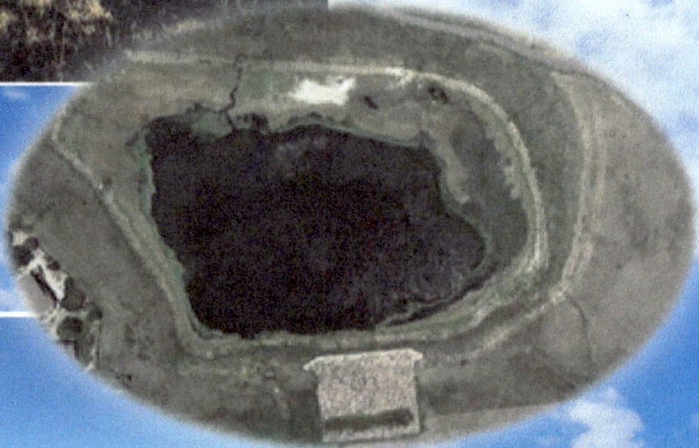

Universally sized, sorted, type granite boulders are uniformed, stacked, flat edged granite as outer hardest shell component as impact design. Dam mortar grooves show it worked by rigid hands, big knuckles and hooved feet. There's no creeks, rivers to dam as found in the study, regardless of size or the world locations.

I believe e the Jamestown, PA flood of 1899 was the cause of one of these lakes collapsing from an 'Earth made dam.'.

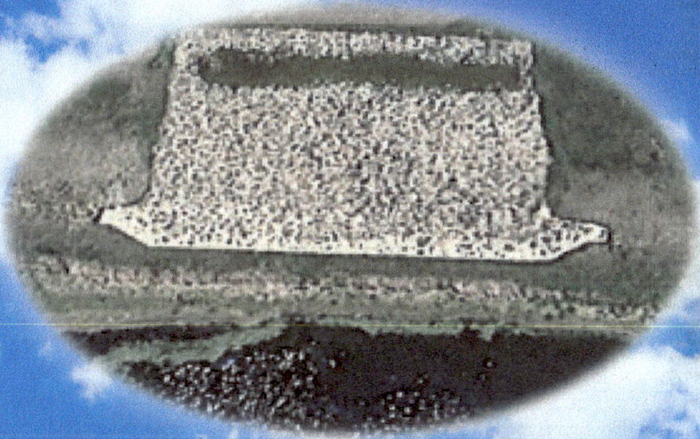

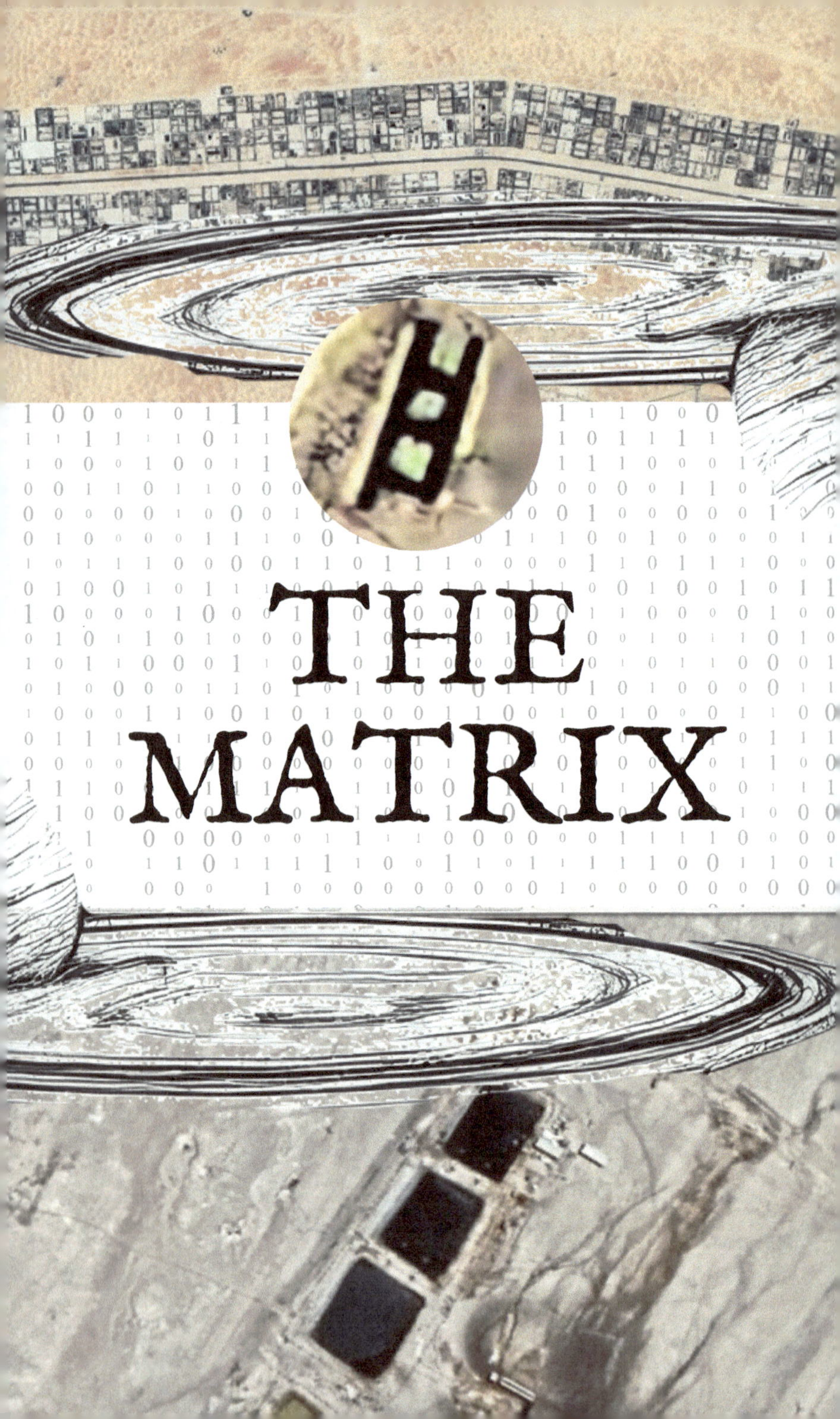
THE
MATRIX

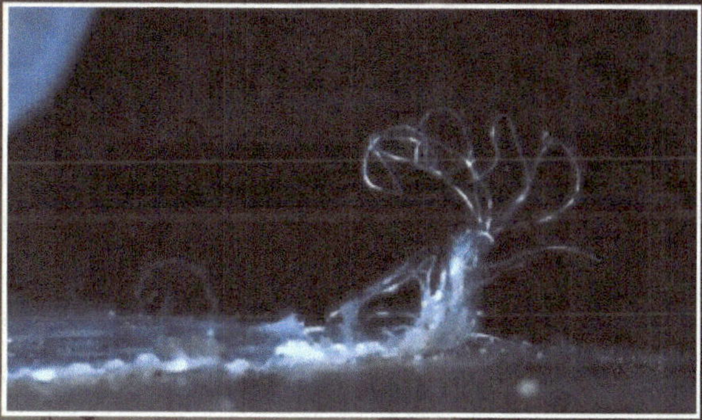

3-cm. rocket grows from consuming carbon from tape strip on microscope stand and its still growing today.

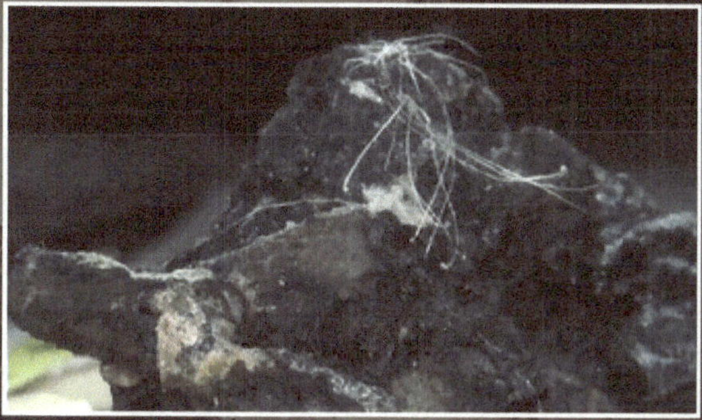

Aug. 1, 2021: T-Minus - 2 cm rocket life appears minutes after I cracked open a black lacquer matrix rock. This species form proved to be highly congruent feature throughout research.

I routinely find high-luster, dense, translucent crystalline rocks with complex geometric patterns, colorations, and varying chemical signatures.

Another complete specimen from within itself, fragments demonstrate the characteristics of continuum and exponentiality.

**Specimen configurations
andmaterials are unique
and easilyfound before
core researchstarts.**

**I can spot them
everywhere Iwalk and fish
once I know theform.**

**The two specimens in
theupper left corner
revealspecies segmentation
andstacking
characteristics.**

It's not a key but a rocket.

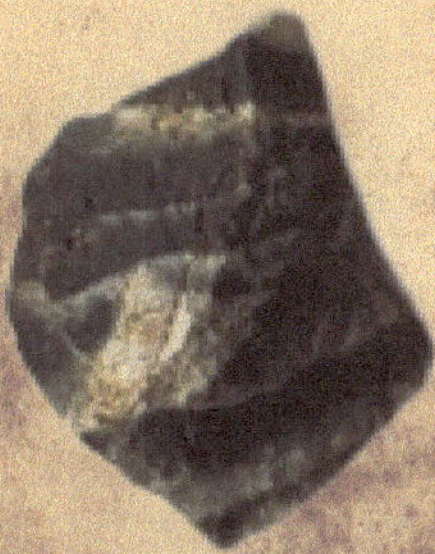

'GREEN ICE' ORGANIC ALIEN CRYSTAL

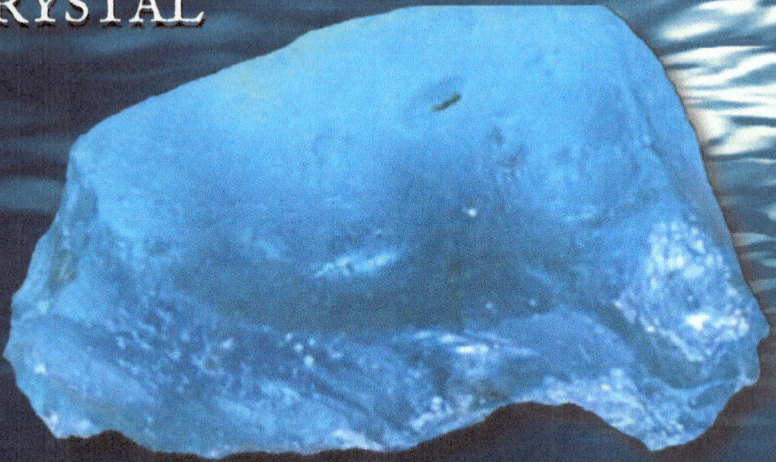

- 10 grams
- 2.5 specific gravity
- 6.9 ~ hardness
- Aquamarine luster
- Extreme water activation properties
- Translucent, organic gas phase, microbubbling, and rapid recrystallization

SEM -

- 48 % carbon
- 28% oxygen
- 20 % nitrogen
- 3% silica
- 1% sodium

Aquamarine colored gem with non-natural formed curved edges, irregular facets with intricate internal crystallization.

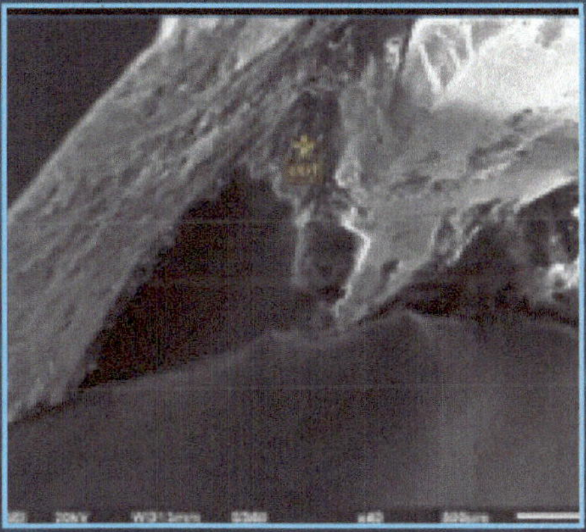

SEM - Scanning Electron Microscopy lends the critical data points in a scored fashion that proves the carbon and water signatures of the various and very different hand samples.

Looking through the water saturated crystal revealed more clues.

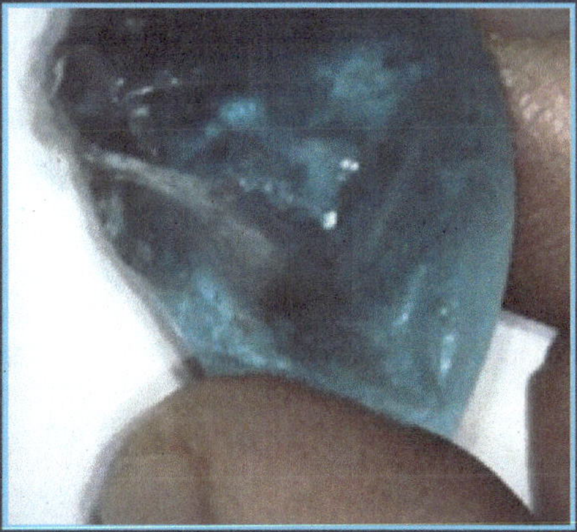

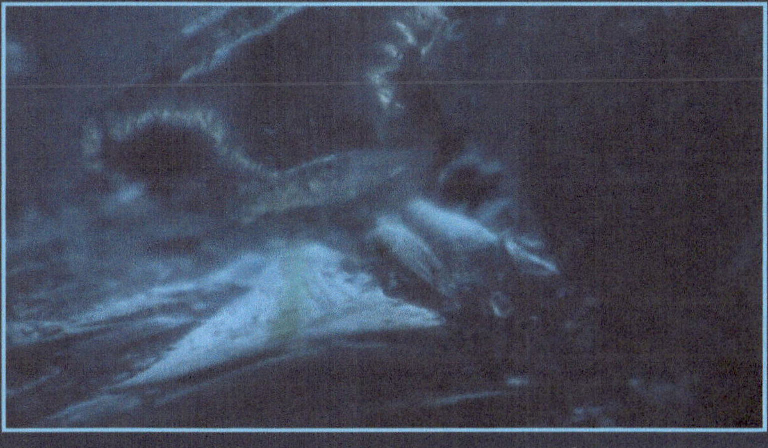

The complex, organic forms within the crystal was like finding a whole new universe and I couldn't take enough pictures of this stone.

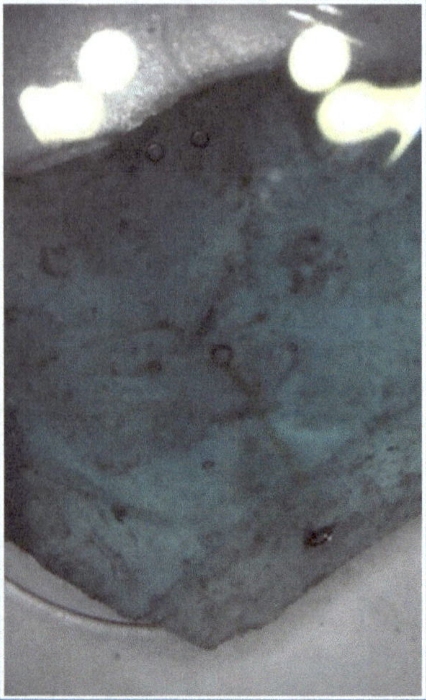

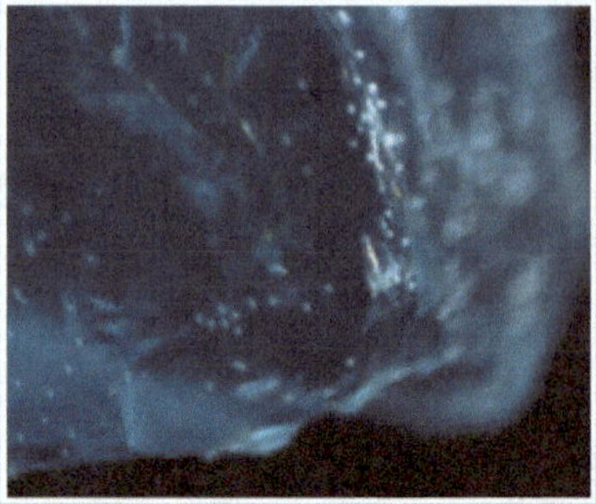

Thousands of nano bubbles appeared inside crystal while submersed in H2 0 then the droplets dried and crystallized to form observable quantum alien nature. It was truly the scenes of scary alien movie that I captured on video.

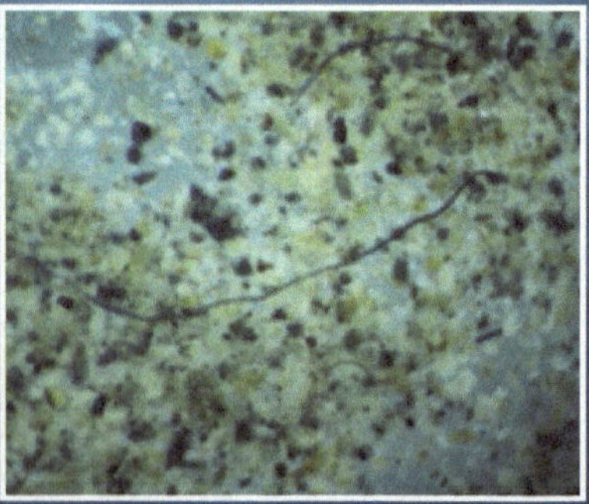

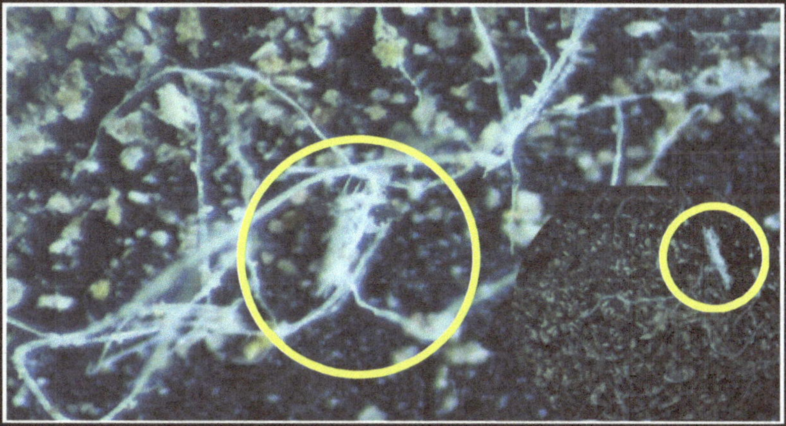

15 minutes later, alien life mineralized, activated out of a water droplet. This is what the String Theory means and looks like to a geologist.

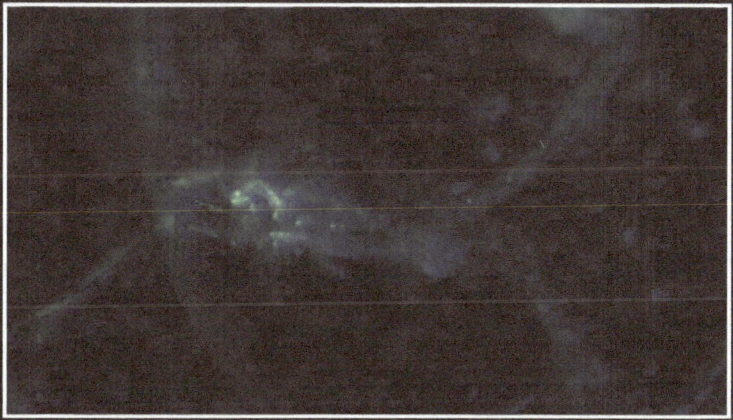

'Strings of Life' is on YouTube.

Finding and observing alternative life forms and understanding their purpose for the Earth is at the core of the research. Every day, there's more evidence to resolve in this story.

Tesonite – Alien Superalloy Discovery

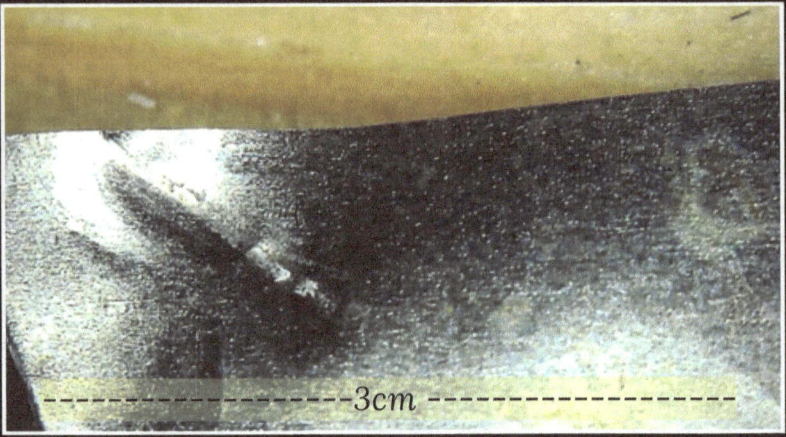

- Produced by a 5lb. hammer strike on a rocket stage that landed on the CO plains.
- SEM: Nickel 34%, Iron 21%, Carbon 18%, Cobalt 14%, Oxygen 6%, Nitrogen 3%.
- 1 cm x 3 cm = % ratio of massive Greenland alloying crystals in the next slides.
- Hyper-light weight, with extreme heat resistant properties assumption.

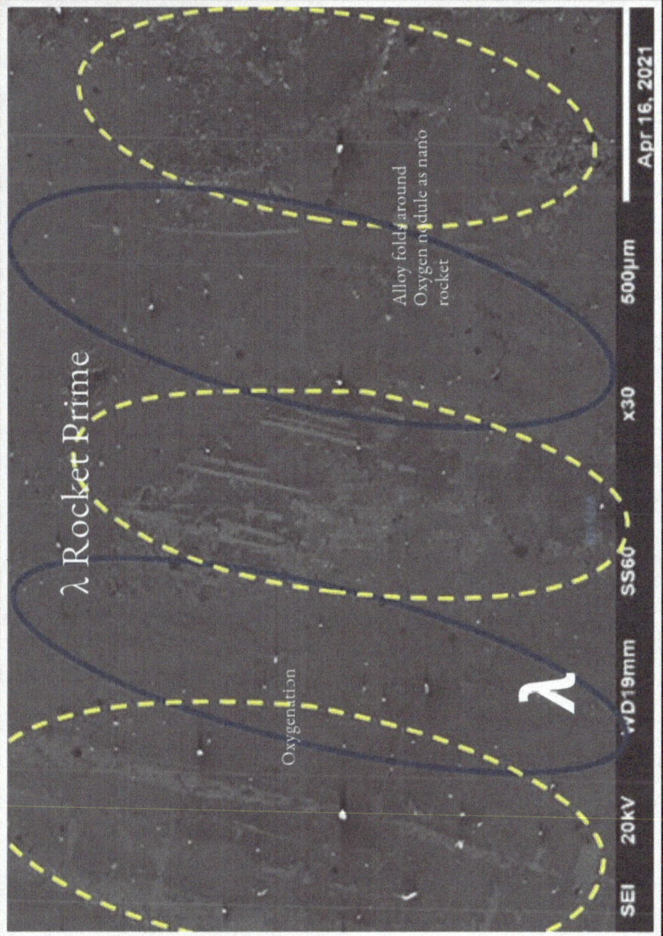

Sothern Greenland - Multicolored super alloy crystals exhibit the exact 3:1 size ratio as the Tesonite superalloy.

I've seen some of this coloration to some extent on the Western plains, but the contrasting perspective to study such large, aligned crystals will provide substantial data to develop a new computational Earth model for scientists around the world. The crystal grain boundaries are the foundational signature of the advanced technology.

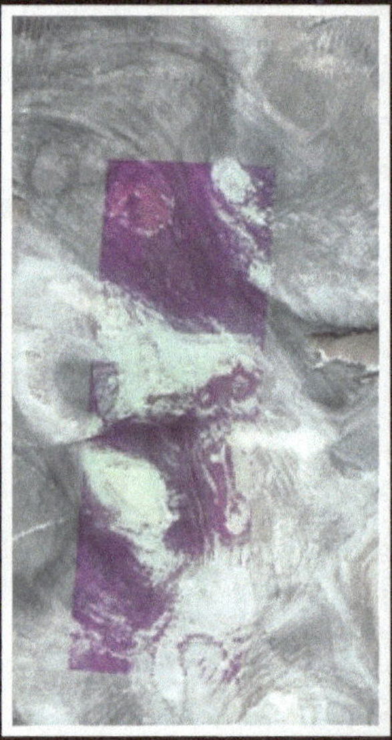

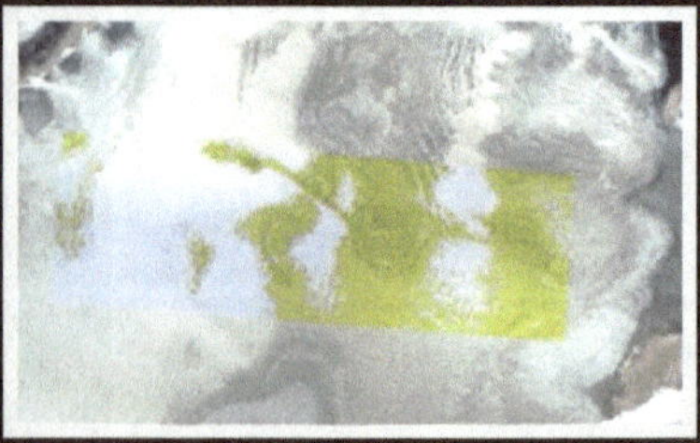

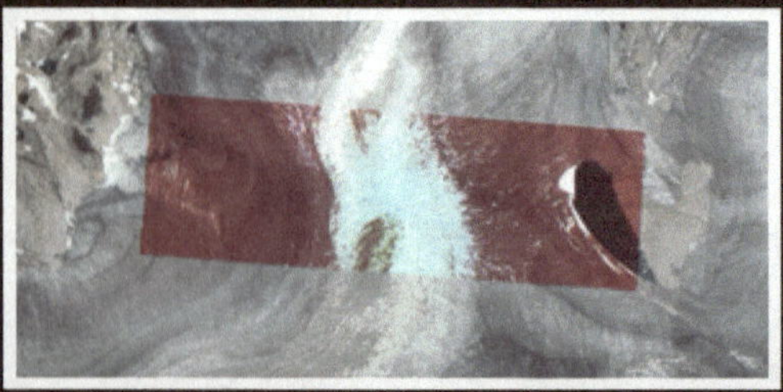

Not your typical mountain cabin in the Wind Rivers but an ET crystal palace built for a long, long, long stay...

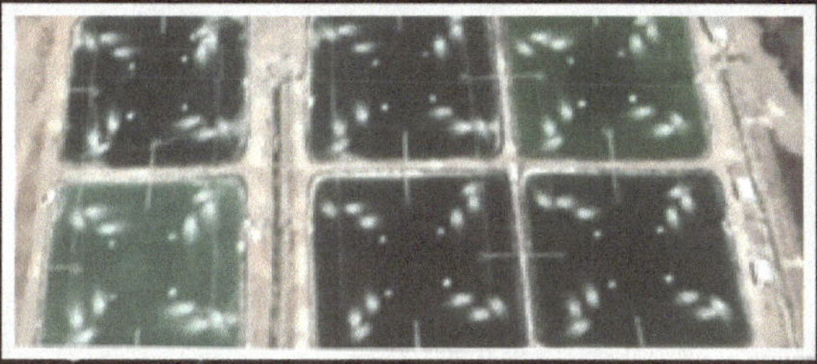

The system interactions, reactions of the alien matrix arecaptured in these 50 ft cells on an ocean beach.
The aquariums were early mind -blowing mysteries in the research that demonstrated the species exponentiality.

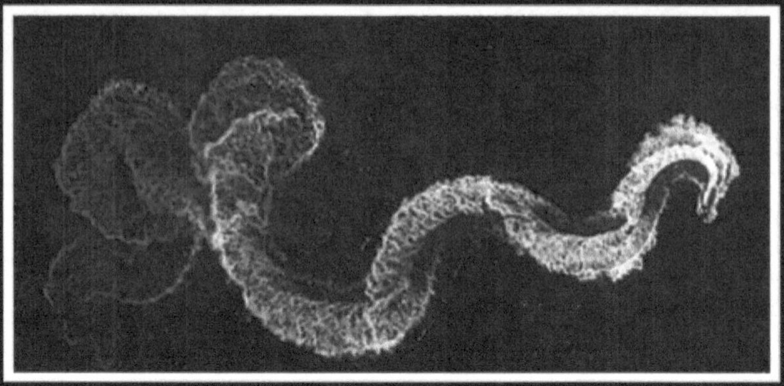

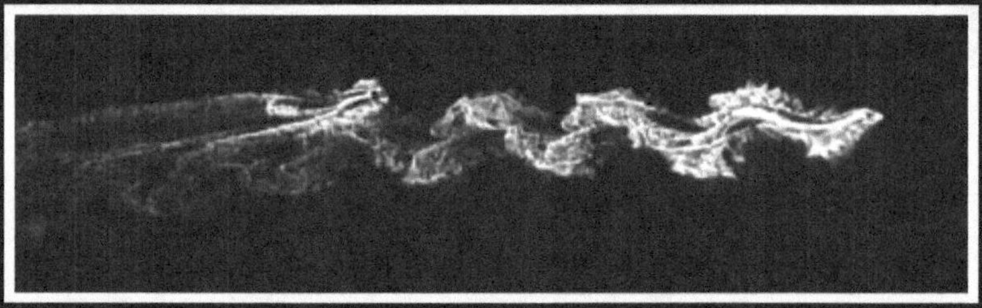

Satellite photographyhas revolutionized theresearch so greatly thesurface images are anendless library ofartifacts of the earthpuzzle.

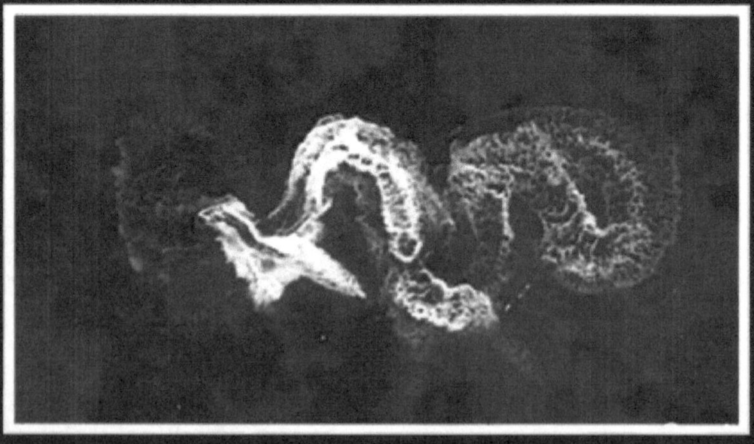

Mediterranean Alien Armada

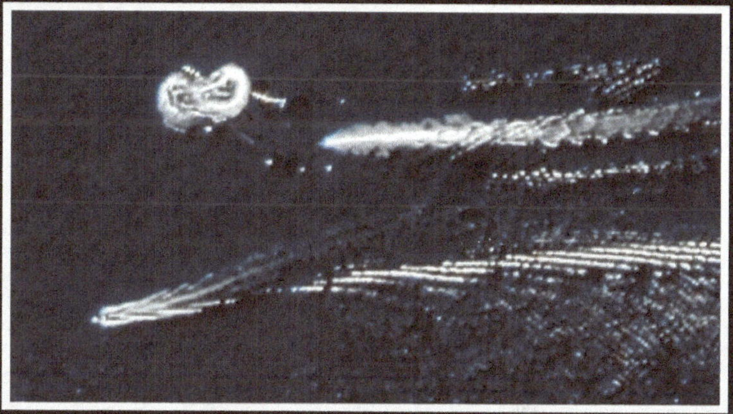

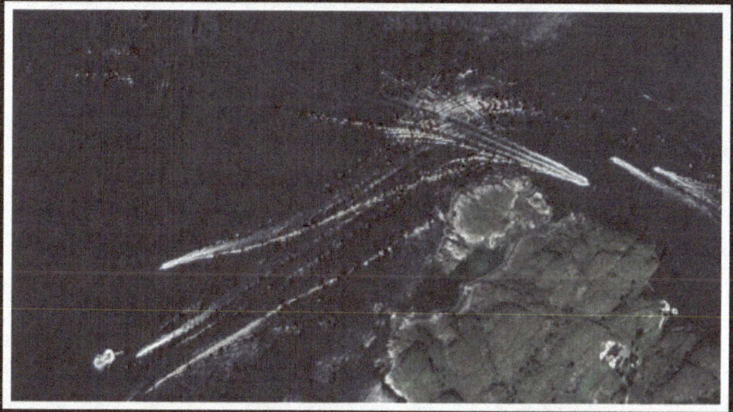

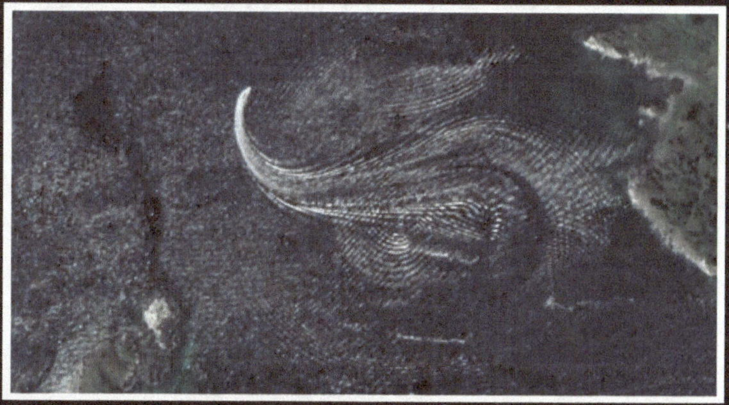

SE Utah – Brilliant grain boundaries are perfected through advanced condensed matter physics and cosmic time.

Middle East

Middle East

Pacific-Rim

Pacific-Rim

Species canvases the mountains of Wyoming.

SCIENCE

The winter and spring of 2020 were when the curiosity trees were planted in my mind through constant and fantastic dreams of traveling to new outer space worlds. I knew and anticipated the potential freedoms of deep physical and emotional observation qualities equipping me to really explore a new world - Earth.

One fine day in March, I woke up with a renewed fervor for space and Einstein. Then, after the second trip to the Winds, it was Max Plank, Boyles, Newton and Wagoner in tectonics as I found a renewed passion for Science, geology, exploration and life. I realized that I couldn't contribute the mathematical applications like the greatest, but I knew my acute field prowess would equip me with new mental awareness prime for a fresh spirit bigger than all the outdoor survival missions casting for lunker trout. The Science chapter assimilates and illuminates the magnificent complexities of integrated science, technology, art, math and engineering in a manifesto of this overarching species.

Crystals of Infinity and electromagnetic pulses

I feel like a rocket flying through space, tunneling each bright, distant object to the next, expecting another portal of discovery.

The research transitioned into a multi-science endeavor when there was enough anatomical data amassed to produce a glossary to term, starting with a new science taxonomy foundations of Einthroplogy, Exterranium, Exterrangia, and Yellowstone. AI - Alien Intelligence.

The chapter will help explain some of the apparent processes given the redundant clues of landings, rocket geology, and colored elements with a quantum material science framework. The entirety makes for a live kaleidoscope of marvelous, out-of-this-world applications, a sub-particle nature replicated everywhere, even on Mars. From millimicron particles to the earth's circumference in thousands of miles that I've found, this is the Carbon species, the life originating from an outer galaxy - living, breathing air and a living land-producing mass. It's how why the continents spread and the climate changes; it's the Earth's roots and branches to anthropology, archeology, biology, zoology, and others explained geologically as a continuum of life matter.

Advanced, multi-science application of condensed matter, physical structures seen in satellite warping, energized rocket superalloy particle arrangement.

Fantastically energized, warping bridges, charging, and altering water are key science pieces illustrating how species control the Earth's processes and resources.

Miles of sand waves altered with an advanced electro magnetic technology a billion years old. The science results reveal energy exponentiality.

Cropped pic below is rod stem intersection.

100% pure, eons aged, powdered molecular dolomite is mined, processed, from its deep previous rocket cycle to re-build same body part in a continuum nature. Top picture shows replication design of Antarctica.

The superalloy manifests from complexities of refraction casting, electro-slag welding, spark plasma sintering, induction and capacitor discharge, powder-metallurgy, ballistics, propulsion, weaponry and eddy currents. These are only of the few recognized and deduced multi-processes within the science realm to date. With the value of time mass and gravity and quantum astrophysics, the museum of Earth is open.

Large, open cavity rocket molds show unmistakable Earth features. With this work, I believe the species built, and connected the Great Pyramids to conduct, transport energy and accounts for most if not all, the mysteriously, vast and complex archeological sites. Aliens are not ancient but extant beings.

Technology seen in transition as the reaction with air creates visible, stacked rocket after reaches it critical temperature, pressure with a specific gravity triggering landing.

Condensed matter physics of electro-slag welding, plasma spark sintering, and powder metallurgy aredeterminable, repeatable technologies with distinct reactions and results.

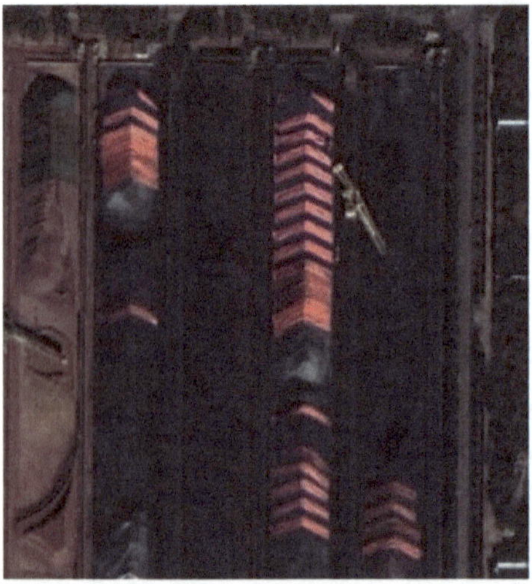

Complex material science of sintering powder metallurgy, annealing in assembly line processing to scale output.

I excitedly turned around to the left side of highway to video these 600' Exterranamine cliffs of the dynamic super matter layers reflecting off water creating the symmetrical beauty of this alternative nature.

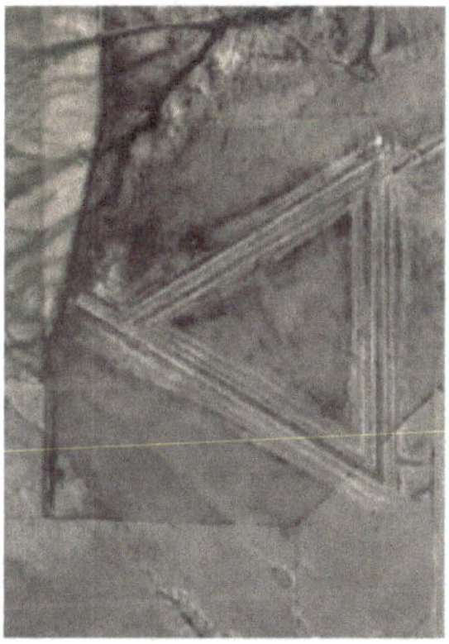

From the Plains of CO to unknown locations around the world, the surface formations are abundant and obvious.

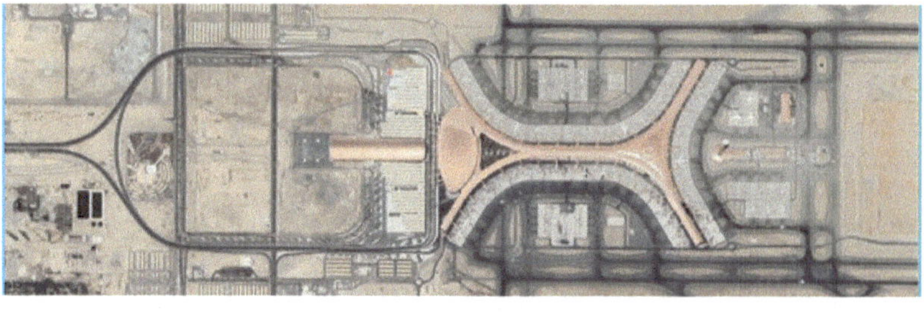

Ballistics blueprint of firing chamber

Wyoming produces the most CO2 than any US state. With such high concentration levels, the state, the country, and the world are a quantum nature breeding ground.

Tunisia; 5.5 mi. species phase replication.

Wind River Basin; March 2024 image. Over four years the organic alloy phase is more pronounced and expanding

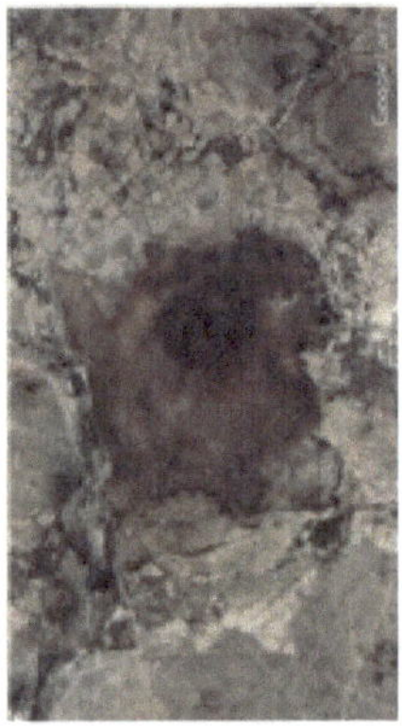

S. China, species phase replication.

Gros Vente Range, Wyoming. Transformational discovery of the 9 - fold-axis, the extreme compression and the uncanny expansion.

The preserved scale of the expansion provided the details to cut out and reassemble rockets to refold the mountain range and the conjoined hills and valleys.

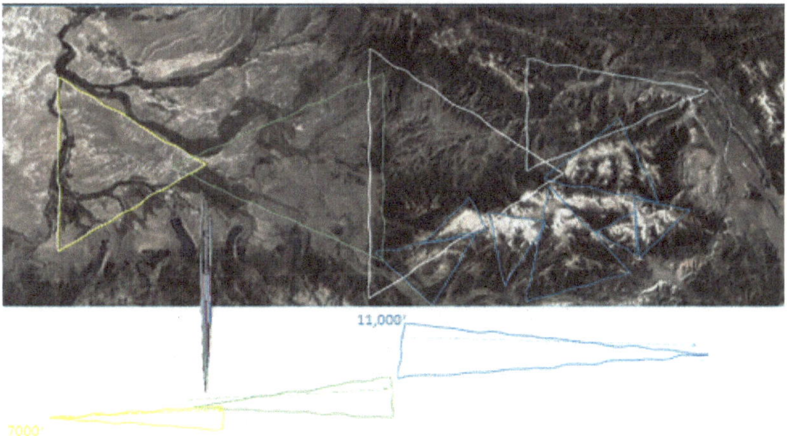

Fold Exhibit

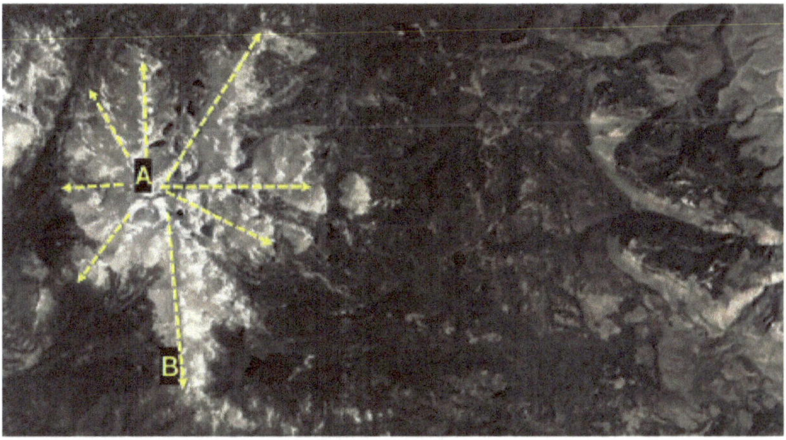

Southern Winds, WY. range rocket anatomy- covers ~ 12 mi^2 core structure.

A - Primary fold axis, direction.

B -Secondary explosion.

Species geomechanics are integrated into coherent footprint that defines its hibernation and exponentially process.

The speed, order, compaction ratio, dimensions, weight, water amount, axis angle is the formula designed over billions of years.

Antarctica always stays connected to SA and then flips over to rebuild the bowl shell as a continuum structure.

The volumetric surface expansion ratio object-to-earth land coverage is estimated to be over 1000%, compounding itself indefinitely.

The world comprises 71% ocean and 29% land, but they are all the same resource.

This seems to be the Super Compaction application found in the geological record.

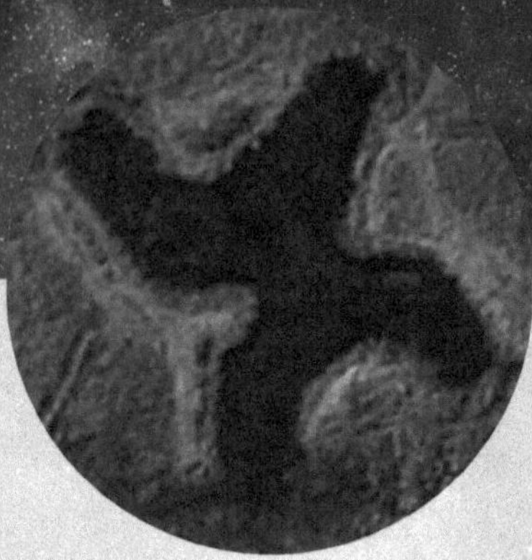

THE PLAN

The grand scheme of the galaxy was, is to exploit primordial Earth by an outer space life written with the stunning records of a massive rocket creating the Earth and Moon. Evidence also strongly suggests that the maiden vessel carried and flooded Earth with the original ocean water, a perfectly blended fresh water source super compacted and designed for an organic rocket that metabolizes water, fueling deep space travel to and from Earth.

Scenes in the chapter unveil the alien blueprints that are illustrated, correlated to handheld lab rockets to ones over billions of years old, in all regions of the world and explained with redundant details.

Comprehending the magnitude of the species' pervasiveness explores advanced STEM on rocket life as the technology for it is the technology.

The Plan is cohesively congruent and provides the only plausible cause of the Great Unconformity, where over 1.5 billion yrs. of Earth sediment mysteriously unrecorded. It has reappeared as geological remnants in the dramatic surface phase changes mass reconstituted as the Invasion bomb matrix.

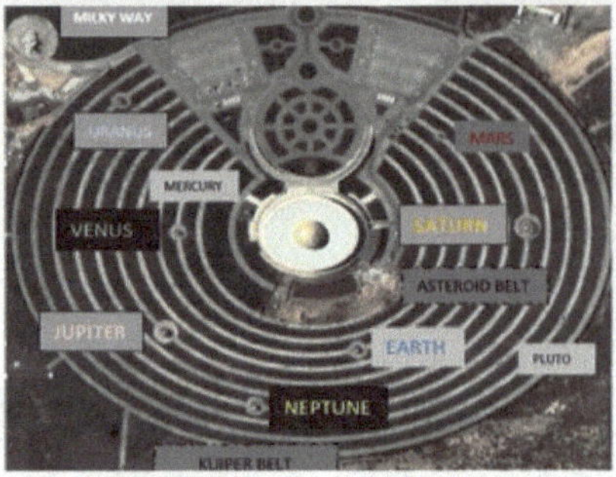

May 2021, Saudi Desert. Textbook, 1800 ft. * 400 ft. elliptical structure of our solar system with planets perfectly sized, colored in their orbits. There is also a conjoined celestial map of the galactical system and the portal that I believe is the species-originating system map legend. This only deepened my understanding there is life way beyond Earth, way out of human's deepest space exploration for now.

Remarkably, this is the same day that Gustav Holst's 'The Planets' was in the public domain to use in my videos. The music has greatly inspired my creativity, and I pay tribute to the artist's fabulous work in Secrets of the Winds.

Author's depiction of interstellar travel

Exterranium: The Earth Event Horizon is the 5200-mile-long galactic rocket that directly hit the hot inner molten metal core, releasing the fresh-water ocean, and starting Earth life.

With a 1500-mile-wide thrust chamber, the so journ must have been scouted long before from a distant system targeting Earth. This super alloy quantum species transported a super compacted amount of fresh water that would cover the entire planet. The steam plume alone from the amount of waterflooding the red-hot planet would have been seen in today's solar system, while the extreme engine heat probably caused a

violent thermodynamic reaction. Front thrusting power ejected a new moon with inner core metal remnants. This set the stage for Earth's new atmosphere, condensing its water vapor to clouds for a greenhouse effect. The entire event with the moon acting as a gravitational ballast caused stronger gravitational pull and created a new orbit pattern given the dramatic increase in earth's (metal + water) mass, density, and specific gravity.

This is the beginning of all of earth's life, as we will come to understand.

The Birth of Planet Earth

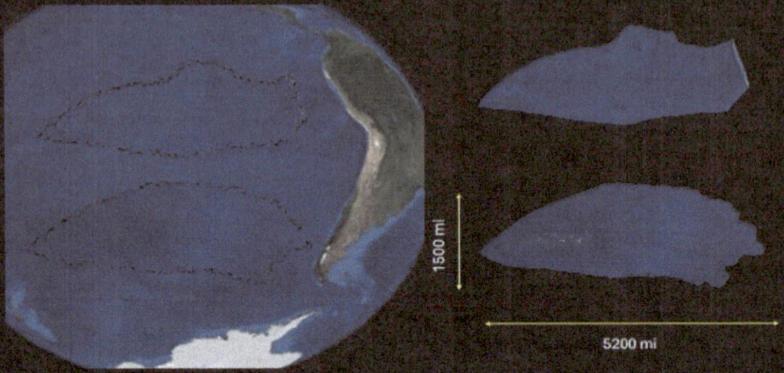

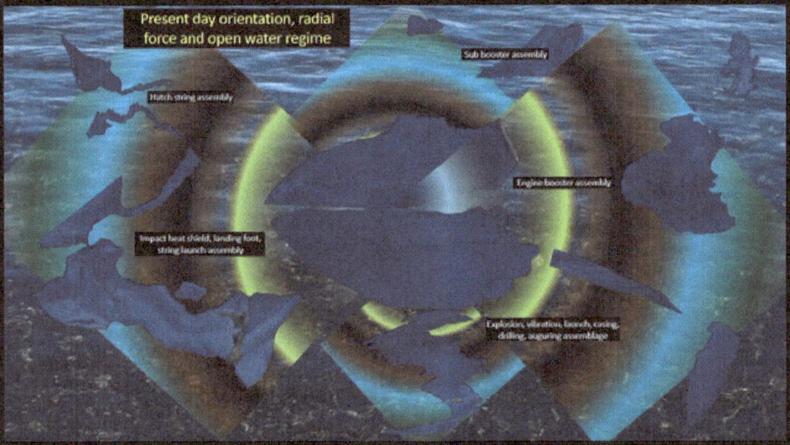

Ocean Earth Event.

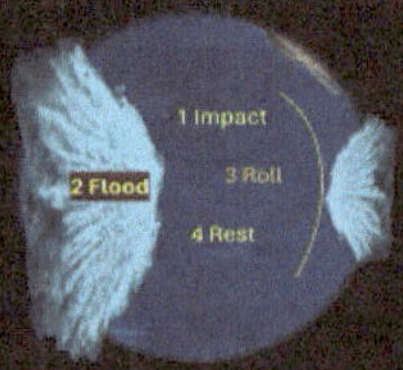

- Impact Event - Hawaiian Islands created at epicenter shown by radiating cracks, force direction.
- Violent flood from super compacted water rocket system.
- Rocket seams explode open, vessel unfolds for thousand of miles.
- Resting inner core of rocket fuselage are the French Polynesian islands.

Conceptual diagram of the rocket cooling and reacting within the new Earth ocean.

Metal folding represents present day coverage of the superalloy rocket seams and inner core vessel. The metal center conveys ultimate control, power, energy and nucleus of the dominate supremely natural force.

Earth turn over and transition phase in a Precambrian Life pre-explosion event concept. There's so much we still don't know still about the depths of the ocean life, but we have now the surface reference data that will reflect the dark, unexplored depths of creation and evolution.

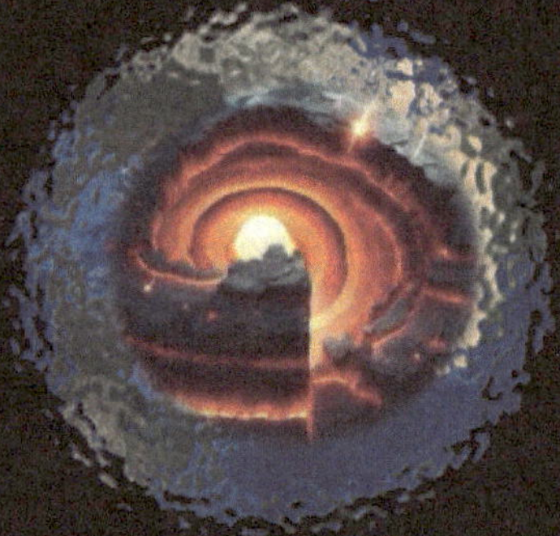

CORE HARVESTING > CO_2 > DENSITY > VOLUME > ROTATION > VEL.,ACCEL. > QUANTUM ENERGY > INVASIONS > SPIN > CYCLOTHERMS > BIOMASS (+/- EVENTS) > MAGNETISM > GRAVITATION > PROPULSION

Event; Exterranium (EXM) 5-2 BYA

Earth is created by the covenant outer spaceship, penetrated deep and deposited, flooded, enveloping the hot core. This was a complex, violent and reactionary episode in which the species first vulcanizes to exude first bacteria flows starting ~ 4BYA. This event also exhibits the geomechanics of drilling, auguring, harvesting, and consumption of Earth ore minerals to blend, manufacturing the future ocean water seeds and bombs. The event causes the subduction and abduction of the outer core of crystalized crustal formations. Induction, conduction energy, and gravity cycles start charging the superalloy and moulding the material for the eternal blueprint. The Great Birth is the Great Conformity.

These are the past scenarios and visions of the Earth before the stabilization of the oceans and the conceptual oxygenation periods.

CORE HARVESTING > CO2 > DENSITY > VOLUME > ROTATION > VEL.,ACCEL. > QUANTUM ENERGY > INVASIONS > SPIN > CYCLOTHERMS > BIOMASS (+/- EVENTS) > MAGNETISM > GRAVITATION > PROPULSION

Exterrangia (EXG) 4 - 2.5 BYA

The seeds of the species may have looked like a glowing pearl, growing under a thick canopy of nutrient rich, ice-cold water. This is earliest Earth ocean aquarium period as oxygen levels increased while the layers start to stabilize and form deep resource processing channels. As the weight of the Earth increases, rotation changed positioning Yellowstone exploding at present day equator utilizing the gravity for optimal spread. The period started the first snowball Earth where the electro- magnetically charges the Earth system. Photosynthesis creates richer diversity for water world earth. Energy cycles stage and produce a diversity ocean life. Subduction core, consumed and recycled sediments to protein chain amino lead to complex life being to consuming itself without and sedimentary stratification until the Cambrian Explosion @ ~ 5$0MYA.

The species form is the land mass model that replicates abundantly in size and expanse.

Earth hatch explosions and the release of Alien Earth Nymphs that formed under extreme pressure and compaction at the Precambrian surface show the story.

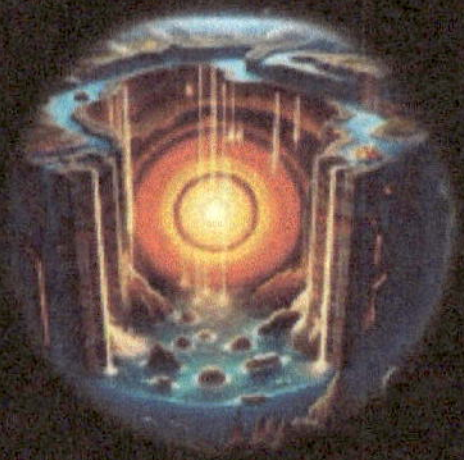

Yellowstone @1.5 BYA

The Yellowstone explosion was the planned process event of the main rocket event radiating from the deep super compacted matter within Earth estimated 1500-2500 miles deep. Post EXG, the Earth had rotated to the equatorial present day Galapagos Islands and 1000 miles drifting northward the spread distance. The explosion was so massive, the reaction is preserved and traced in all world's continents and oceans.

The Yellowstone even created the major fresh-water surface sources along with the predictable geysers to the volcanics which are the routine but complex geological cycles and the true nature of the mechanical rocket species.

ENERGY > CORE HARVESTING > CO2 > DENSITY > VOLUME > ROTATION > VELOCITY > ACCELERATION > INVASIONS > SPIN > CYCLOTHERMS > BIOMASS DIVERSIFICATION, MASS EVENTS > MAGNETISM > GRAVITATION, MECHANIZATION > PROPULSION > CLIMATE EVENTS

*geological attributes over time

July 2024 - Returning to the Totem Rock discovery from 4 years ago, It was conclusive that the cliff face shows the twisting, lateral folding and vertical coloration sequence of the Yellowstone pipe during the Precambrian. The Google Earth image below also highlights the extreme geological event. The outcrop is a prime exhibit of the atomic power of the event thrusting billions tons of super compacted, spinning rock 150 miles away from the Yellowstone epicentre.

South of Devils Tower, Wyoming is a massive rocket ship blow hole (left, white) and 6150 diameter hatch TOP (right, green) also known as Green Mountain later explained with X-section, explosion scenario.

Yellowstone was traced to the Galapagos Islands based on the under-water structures and crustal drift to current day. The islands successive and stacked craters reveal the bomb pipe explosion seen in the connecting the land mass and the following expansion process.
Birth of the early continents; Exterrangia - is the new definition and interpretation of the earliest land mass, and the origin of Pangea.

1000 mi. rocket species between UK and Iceland is largest intact and earliest in age model. It's built to propel and replicate and the model bomb for the Galapagos, thus Yellowstone.

South of Devils Tower, Wyoming is a blow hatch (left) and 6150 diameter hatch plug (right) also known as Green Mountain. Circled are the oldest inhabitants of Earth. After explosive launch, residue is seen light as colored surface matrix strewn for 12 miles.

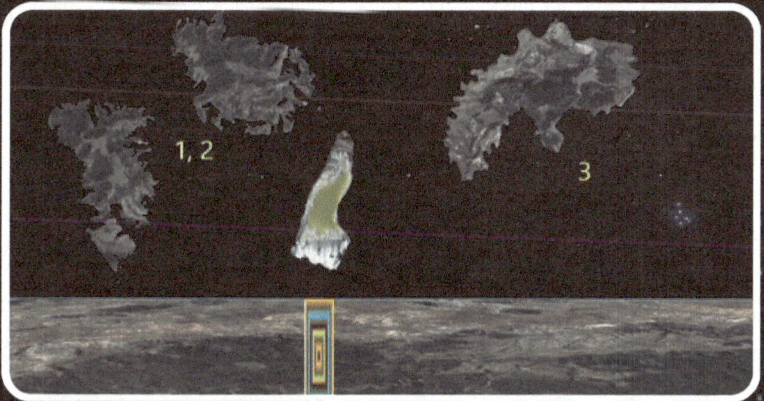

Precambrian rocket hatch explosion and release of alien Earth nymphs 1,2,3 that formed under extreme pressure and super compaction over billions of years. Rocket is from Bolivia to show load concept and launch to sprawl design.

Exterran Infinitas Giganteus (EIG)

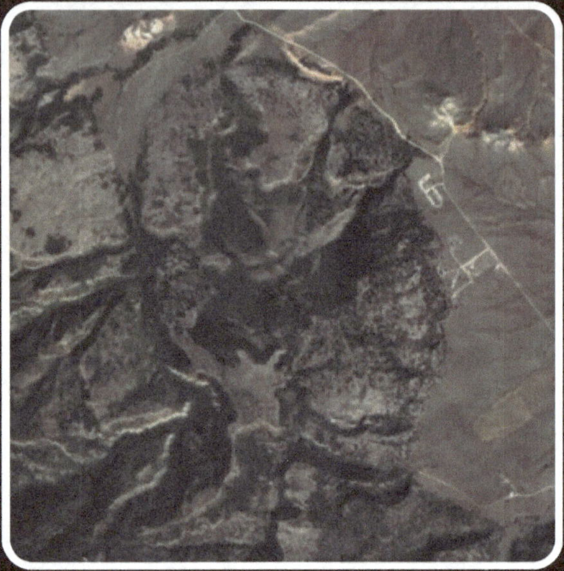

Near the Green Mtn hatch hole X - section in Eastern WY, this alien profiling measured the heads that are 1.5 miles long and they exhibit external lungs, jaws, eye ridges. These giants were borne out of time when the alien matrix was buried deep hundreds of millions ago year growing deep under high temperature and pressurized chambers.

Giganteus' anatomical features are exact as in the microscope and the bones from EIN. It's a distinct family nomenclature with members varying in extreme sizes, different geological periods with different functions. The impacts on Earth and the effects on human cultures are just being understood.

E. Idaho; this is the biggest heart found to date; 10 mi. long and the location of journal nine (9) chapters symbols. This the only interpretive, communicative site found so I decided to use these to outline the work coherently and cogently.

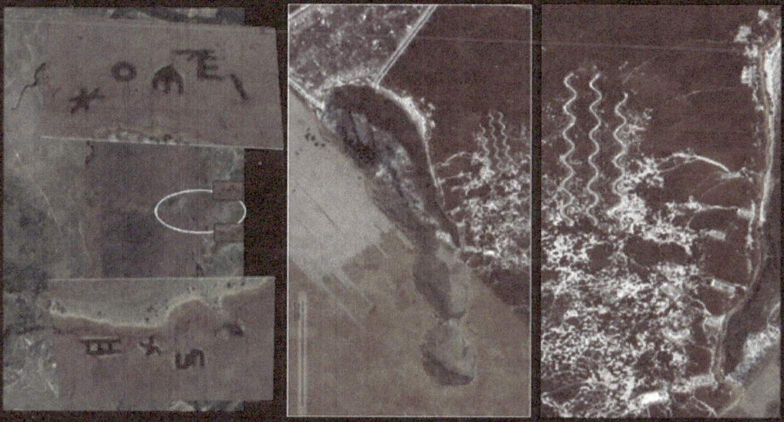

Heart pump channels processing oxygen in white.

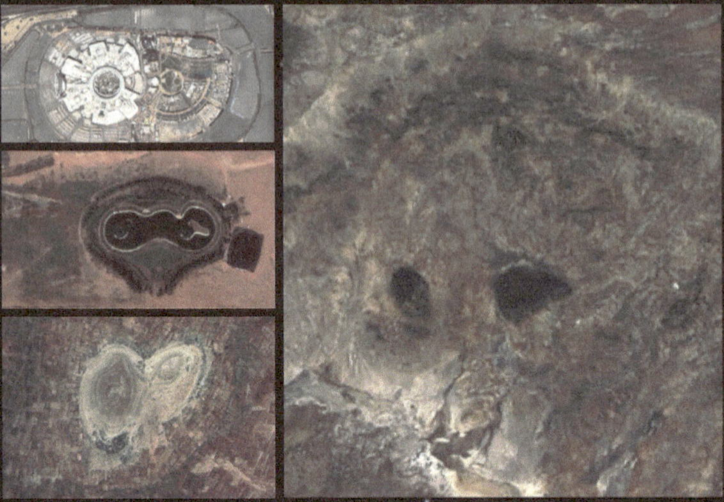

Depending on the geological time of the landing and the available ambient water resources, the eyes bring out the heads, faces, snouts and external lungs of its patterns traceable around the world.

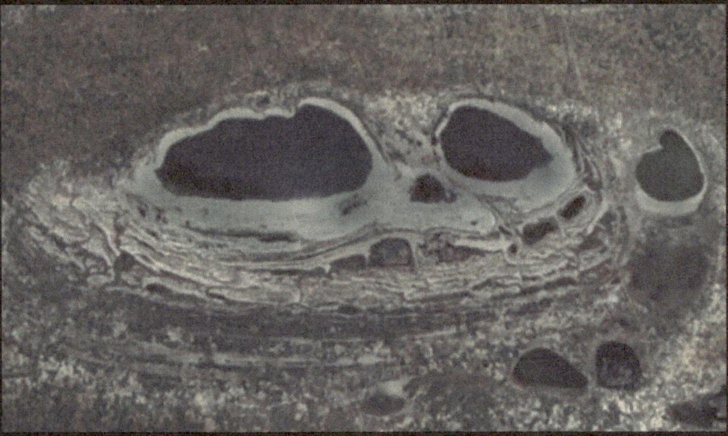

Extraterrestrial beings have always been here and imbued in the earth. The research concludes the discoveries of the family species include creatures like Big Foot, Swamp Creature and other infamously mysterious monsters of modern tales. This is the missing Earthling Mozambique EIG monster head is six miles wide with lake-filled eyes, fangs, hollow, pointed dart tongue and the external segmented lung once connected to the nostril bone.

This was personally the most terrifying alien that I found to date..

Reconstituting, reconstruction of an alien anatomy details the trenching or segmentation of the elemental, mineral concentrations. The properties and the associated colors attribute to hardness, combustibility, electro-chemical reactivation for specific functions.

The Great Salt Lake Utah, April 2023.

The lake is so low and dying that the mineral concentration has fouled the air, and it's become a toxic cloud for the locals. However, the receding shoreline exposed the extensively complex and ancient water manipulation structures and the connected materials science processing apparatus.

I ventured to the remote SW corner, where an outflow is created to feed the alloying rocket in real-time. The trip validated all the research and shows the Great Salt and Utah Lakes are remnant lakes carried from the Yellowstone explosion.

This is the Great Salt Lake River.

I noticed seagulls were only concentrated in an area where there are miles of shoreline that have been manipulated, perfected with berms, deep trenches, graded to harness, and trap lake water to flow OUT of the lake.

I was astounded seeing ~ est. 50cfs H20 outlet feeding the nearby rocket plant process.

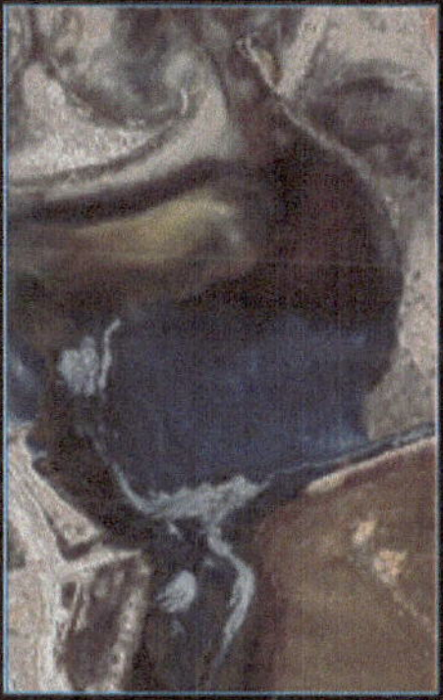

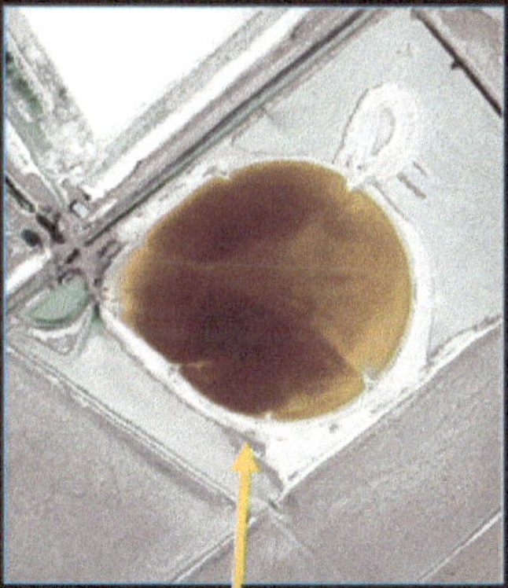

Great Salt Lake, UT, the rocket species is vividly exposed with large scale, colored, reactively forming living components.

The complex organic multi-colored materials science shows the rocket superalloy factory that the GSL water feeds.

The Great Salt Lake sub booster rocket, alloy matrix is still one of my earliest and favorite rocket photos, and it proved to be a critical piece of the story by correlating real time by seeing the alien rocket plant damming the lake water to flow to process metal. The landing block upper left is taking its form.

The Rockies especially UT, expose the beautiful, large-scale macro depositional metallic nature from the Yellowstone Event. The result is also why UT geologically Is such a panacea of brilliant, deep canyons and full of colorful hot beds of trilobite fossils.

Sub-Booster Rocket Global Correlations

Bolivian desert - 4 mi.

Baja coast NA - 650 mi.

Great Salt Lake UT – 13 mi.

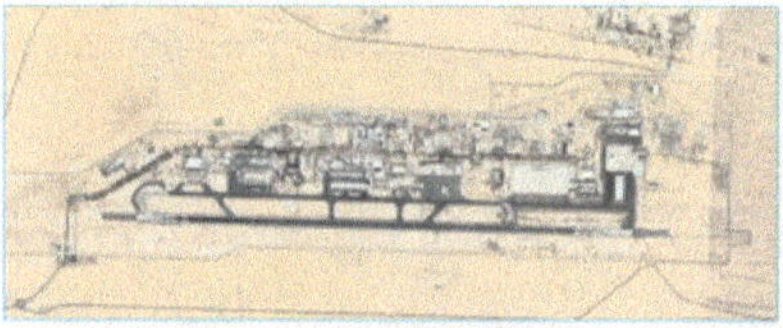

Desert Rocket blueprints will be studied in future journals.

Photo of an atoll rocket transforming into the early superalloy, natural geometric process.

Species recognition in the microscope and satellite is key to solving the big picture.

The results of the advanced technology are a magnificent display of a quantum superalloy nature created over billions of years. The earliest Precambrian animal species may have been more metallic and robotic looking but may have changed with the Earth and the food sources.

Its renders like a croc, ray, shark, eel, fish, and whatever else I can imagine matched to the rocks.

The Primer is the quantum code of the species life strand. This is how, why and what the String Theory application is to a geoscientist.

Primer electro slag molds and relics

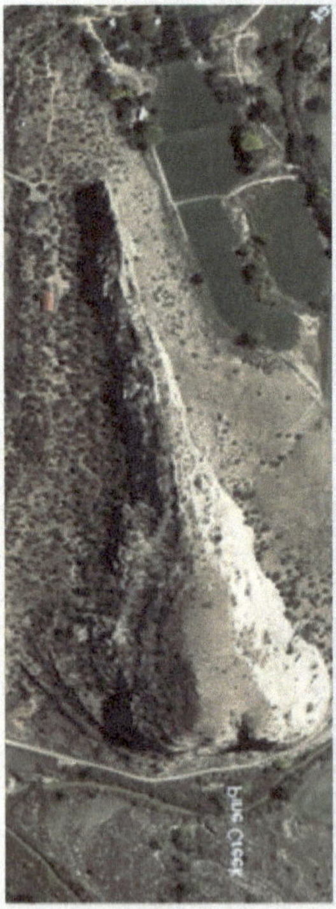

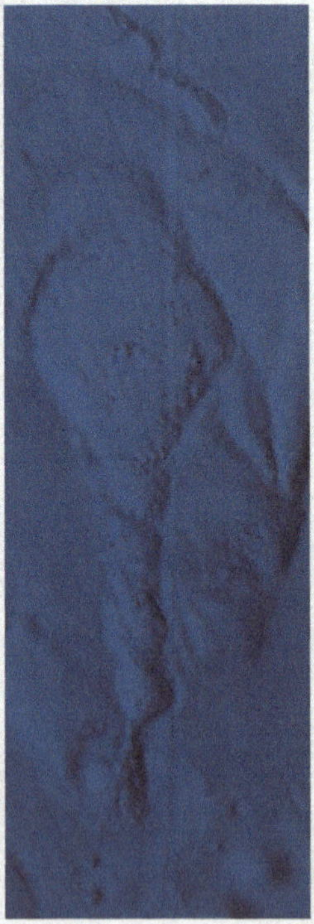

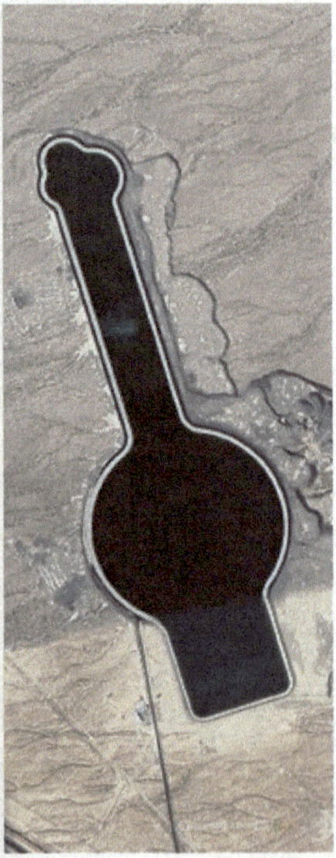

RETURN
BURN

The chapter symbolizes the explosive nature of the alien system. It also provides visual sensations of rocket species' active ignition phases through plasma spark sintering, cloaking, twinning, and landing phases.

Living through the multiple scenes at this stage of the research, I came to the humbling, divinatory facts. I resigned to the immense depth of human oblivion about nature, including myself.

The Exterranamine is a real-time, material-mining and rocket science factory that demonstrates mechanical, combustible, and rocket reconstitution technologies in real time.

Perhaps it's time to explore further with the species to understand the nuances of human interactions better. We need to find out who dares us to step inside and attempt communication.

The plan for the next personal life stage is to blast off in an Earth rocket to explore the lands of an alternative life planet, perhaps the one that NASA has just found.

Mars exploration and future missions could be studied best on Earth, where its deeply engrained technology can be applied to the deepest missions on the planet.

All clear for take off...One of my favorite all time Google Earth photos. It's no more wander why so many pilots, humans witness these flying objects.

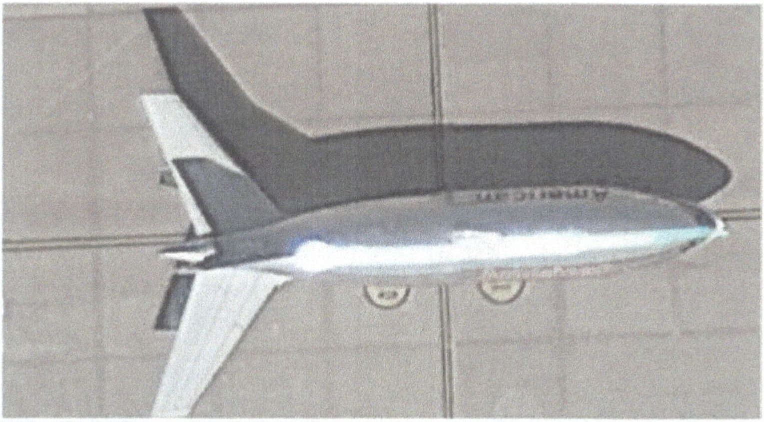

700 ft (230 m) long rocket plasma sparking ignition phase. One of many in this area, these advanced launch stages have only been found and concentrated near the Cobalt Crater discovery.

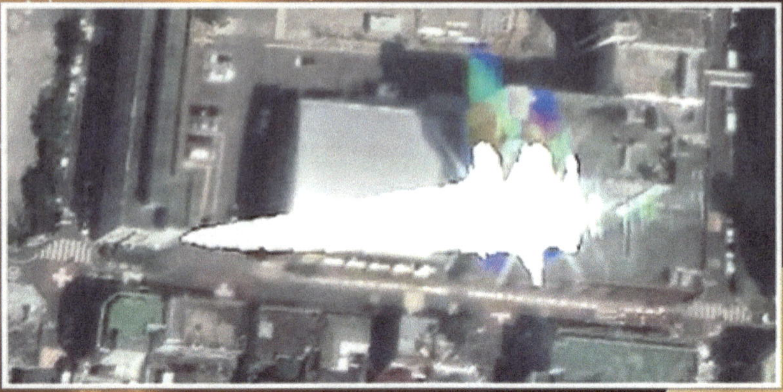

ALIENS ARE THE TECHNOLOGY;
THIS IS ROCKET EARTH.

The rocket plasma factory shows a quantum spectrum of cloning, morphing material disguised as gravel mines, and assorting facilities that are producing advanced alien rockets for blast-off.
This series of photos were rendered only on iPhone 11 and not AI-generated.

The mechanical alien conveyor belt exhibit process is only seen on Google Earth and is not visiblefrom the road in our visible spectrum. But I've never seen any moving belts or loading trucks everbut new huge piles of color-assorted earth.

I couldn't believe my eyes at first, nor did my family, for that matter.

Once visible from the ground, these power lines, can no longer be seen from the road or on satellite.

POWER LINE AND DAM STRUCTURES
<u>EXTERRANAFORMS.</u>

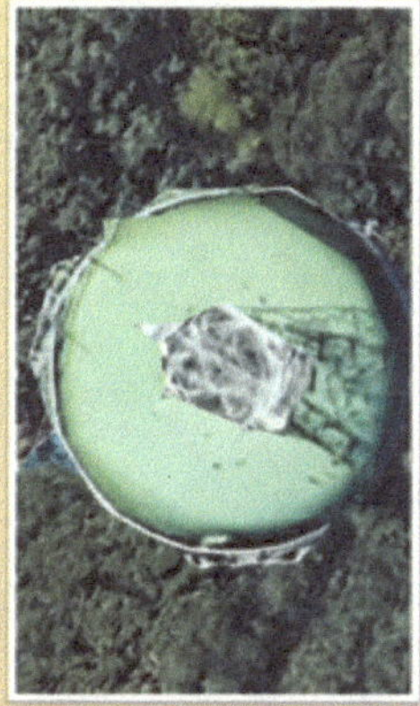

FRONT RANGE COLORADO

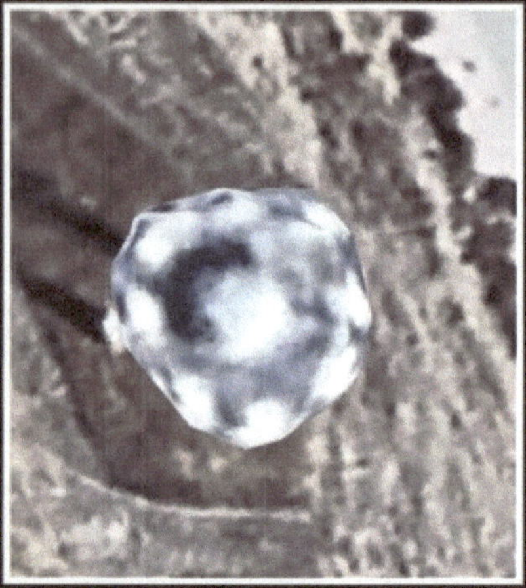

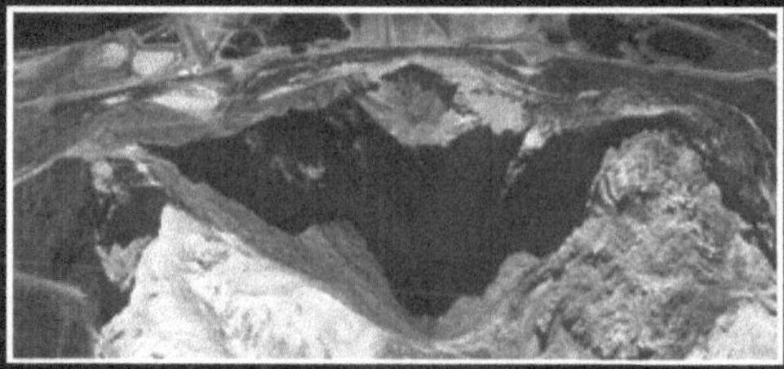

Rocket Mountain Factory, Jones Pass, CO. shows pentagonal structure and hyper-dimensional outcropping.

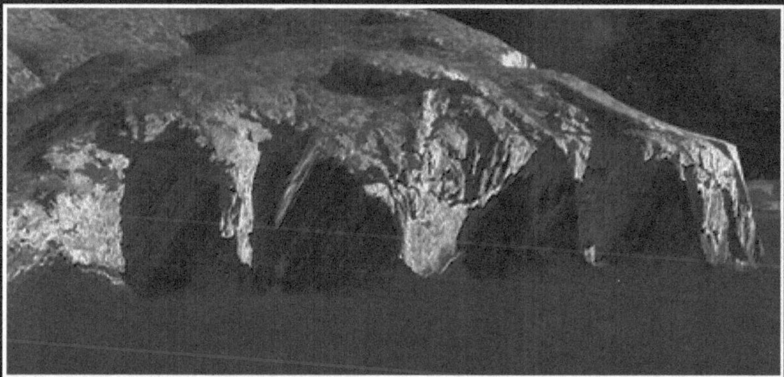

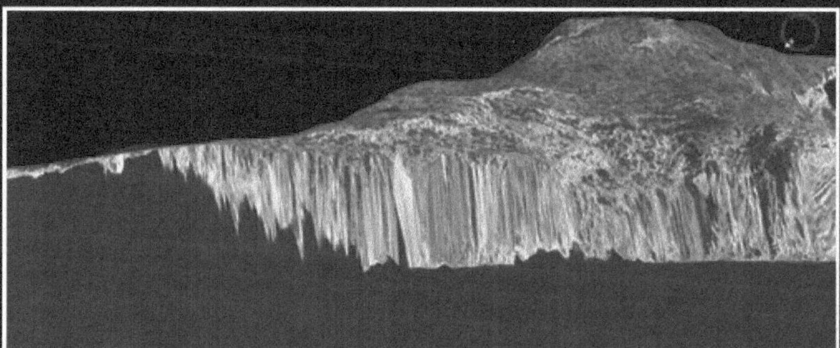

Mediterranean Sea cliffs show the alloy outcropping.

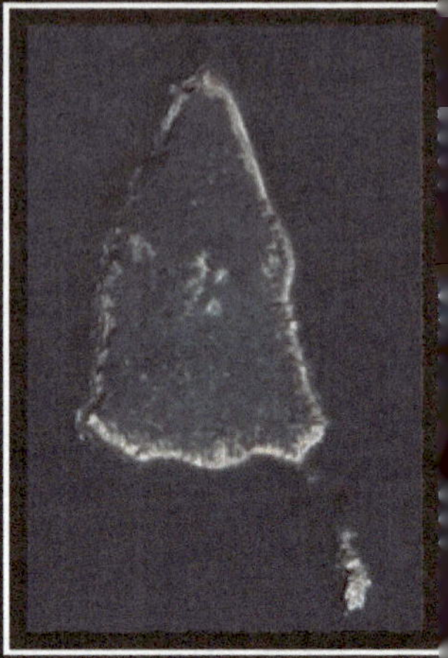

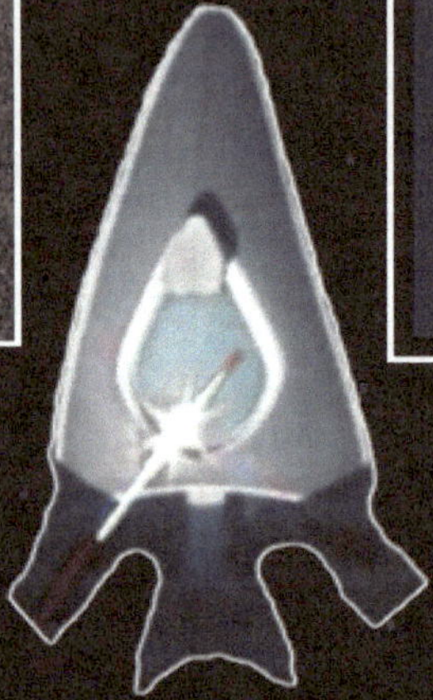

The Great Conformity; the past, present and future, the Earth species returns to reform and launch. The quantum, geo, bio mechanical design perpetuates itself through time.

'Green Ice' 100-micron rocket

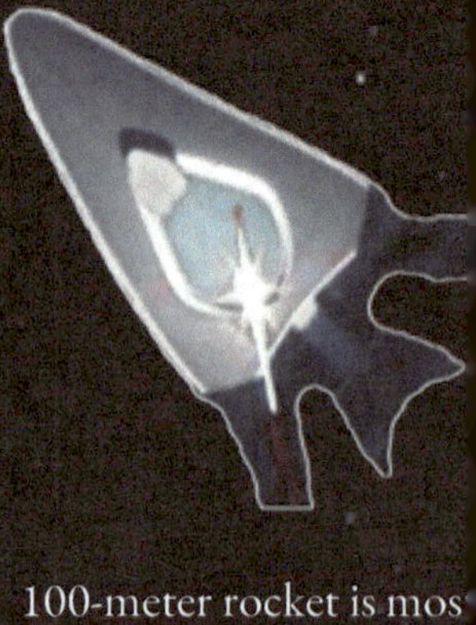

100-meter rocket is mos definedfound to date .

Exterranaforms –Alternative dimensions of nature observed transforming, hidden in human structures and human-like objects that are rendered in plain sight.

700 ft (230 m) long rocket plasma sparking ignition phase. One of many in this area, these advanced
launch stages have only been found and concentrated near the Cobalt Crater discovery.

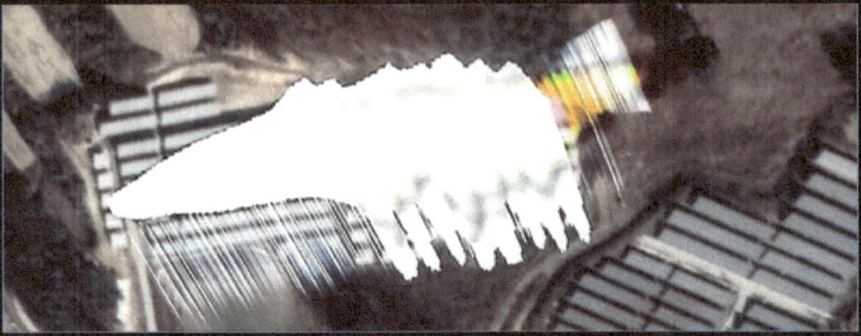

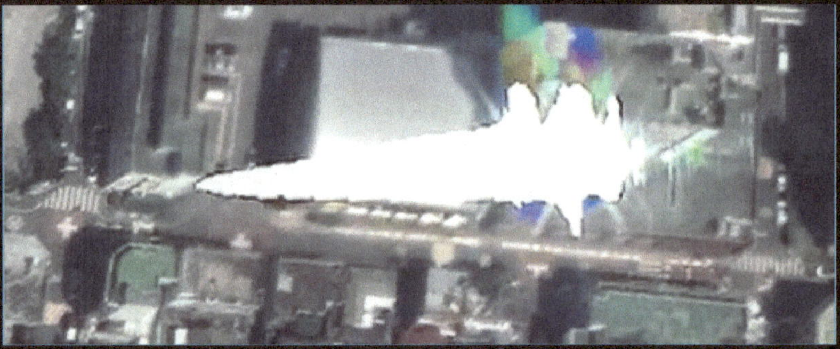

Aliens are the technology; this is Rocket Earth.

Our planet is a geological time continuum hub of an advanced ET species that has been invading, sprawling erupting, and launching for billions of years.

INFINITY

The quest for understanding leads us to find the spark of life on Earth as a singular occurrence, a part of a grand cosmic design.

The theory of panspermia suggests that life's seeds could be widespread throughout the universe, potentially shuttled between planets on the backs of comets, asteroids, and rockets, making the notion of life as a cosmic phenomenon not just a possibility but perhaps a commonality. This idea tantalizes the imagination, proposing that Earth's life may have extraterrestrial origins. In turn, life from Earth could find a foothold in distant worlds.

The search for exoplanets, planets beyond our solar system, has revealed thousands of worlds, some of which reside in the habitable zones of their stars. These are regions where the conditions might be right for life as we know it to exist. These discoveries fuel speculation about the existence of another Earth-like planet, a place where life has emerged and thrived in parallel to our world. The James Webb Space Telescope has even detected potential signs of life in the atmosphere of a distant exoplanet, hinting at the possibility that we are not alone in the universe.

As for living in another time, the relentless march of evolution suggests that humanity will continue to adapt and change over millennia.

While we cannot physically travel back in time, our species may evolve into forms that future generations see as distinctly different from us today. The concept of time travel remains a staple of science fiction, a fascinating idea that challenges our understanding of the fundamental laws of physics.

The distance to these other systems is vast, measured not just in miles or kilometers but in light-years—a unit that conveys the immense scale involved when discussing the cosmos. The nearest stars are tens of trillions of miles away. The exoplanets we've discovered orbit stars hundreds or even thousands of light-years distant from us. The journey to these worlds, if possible, would be the ultimate odyssey. It would span generations and require technologies that are currently beyond our wildest dreams.

Capturing the essence of humanity's eternal search for meaning and connection in the universe is a lofty endeavor. Through scientific inquiry and imaginative exploration, we all seek to understand our place in the vastness of space and the continuum of time. Whether we find our cosmic kin in other worlds or discover that we are indeed unique, the journey of discovery promises to expand our horizons and enrich our collective knowledge. The answers to these questions may not only redefine our understanding of life and existence. The sensation of infinity that I've embraced is a testament to the human spirit's unyielding desire to reach beyond the known, to touch the fabric of the universe and find our reflection in its infinite depth.

THE LITTLE GREEN MEN ARE NOT SO LITTLE AFTER ALL.

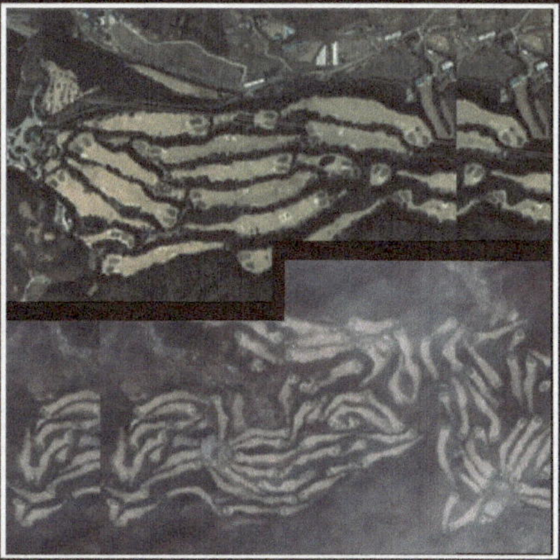

These golf courses looking farms were found specifically concentrated in the Pacific Rim.

Aliens are laid out absorbing sun and developing according to the launch plan.

Compared physical evidence is the undeniable result that's seeneverywhere on Earth.

300-micron photo of the rocket species from first research samples from the Winds.

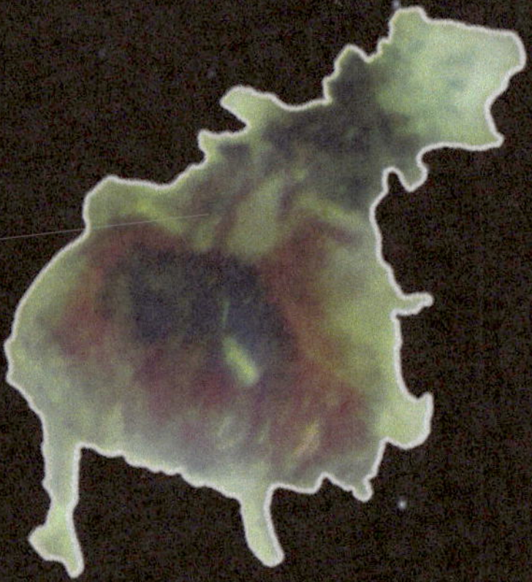

I had to travel all around the world to circle back to Wyoming and see the Yellowstone origin of this index crystal.

Metallic monster exhibits perfect grain boundaries outlining the head and horn features.

Infinite, invisible dimensions exist on Earth hidden in plane sight seen in rockets cloaking and disappearing.

Compared physical evidence is the undeniable result that's seen everywhere on Earth.

The Crystals of Infinity scenes encapsulate the journal's universal theme that we live in astonishing brilliant, quantum physical world, unbounded with imagination to reach beyond human time, in unlimited dimensions and our inner space. The showcase illuminates the superalloy life cycle crystallization process in the purest and oldest life structuring over billions over years, transforming miles of the ocean and coastline.

The top Tesonite crystal (330 m. long) image can't be found on Google Earth again, but I got at least this one.

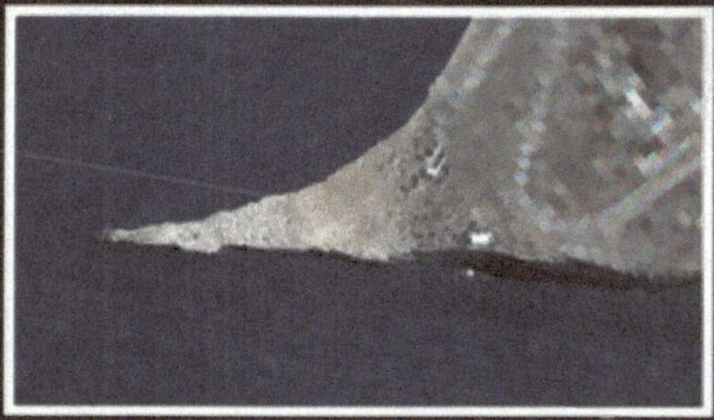

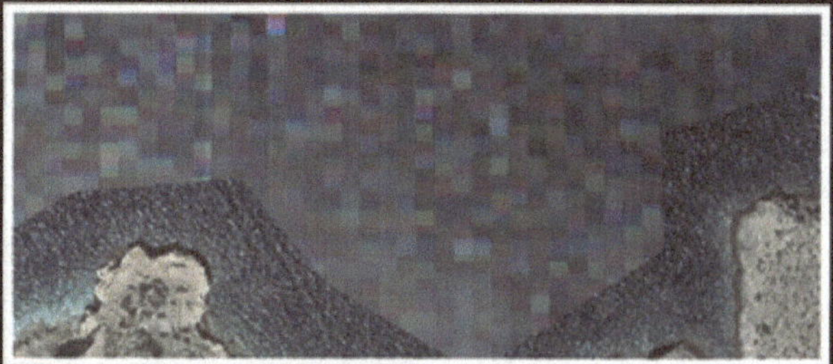

Crystals of Infinity

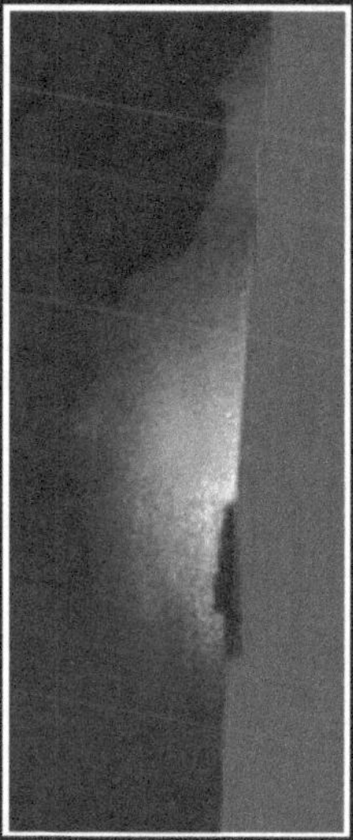
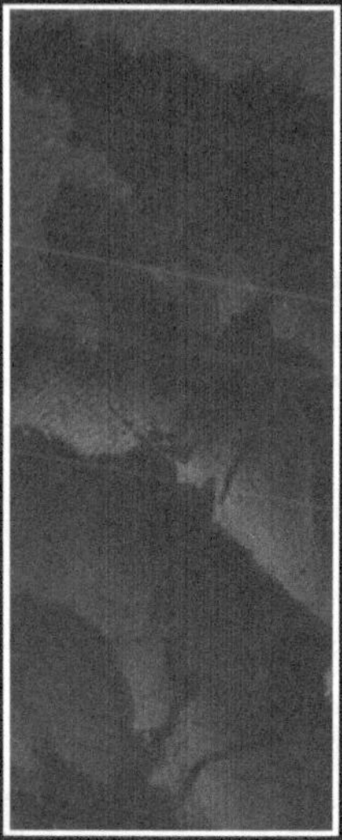

Our world is a connected, multiverse masterpiece of parallel space, time and matter grounded by our lives on Earth.

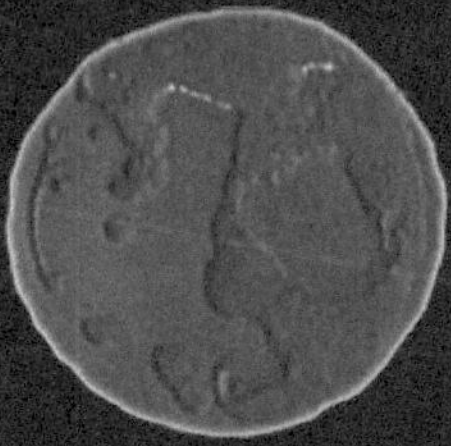
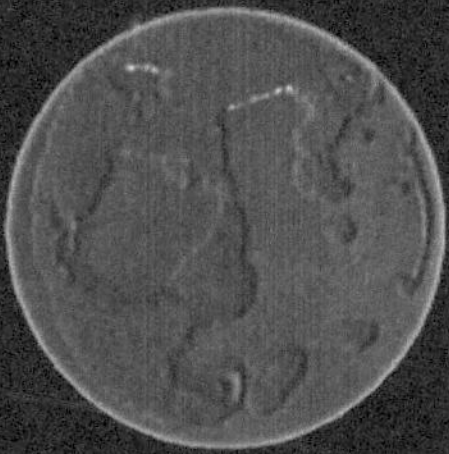

100-meter beach sand mold of world map looks like early Earth infinitely forming itself. The picture changed itself digitally, another uncanny experience during this journey.

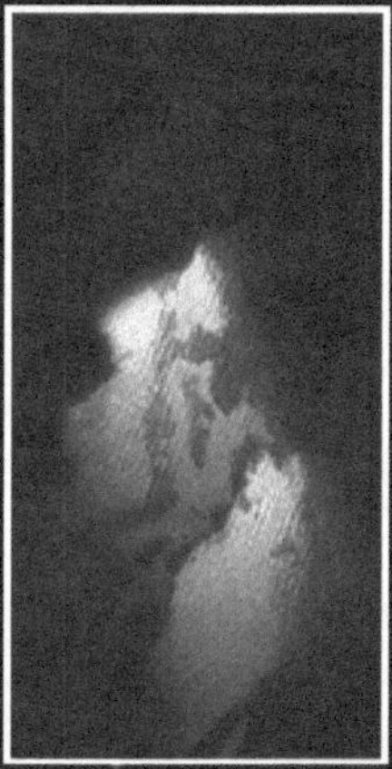
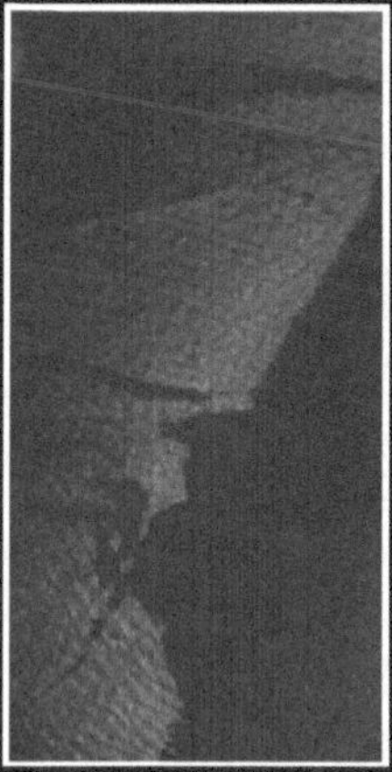

Evidence quadrillion miles away, deep from outer space to the SW Wyoming plains, this is the master species to behold.

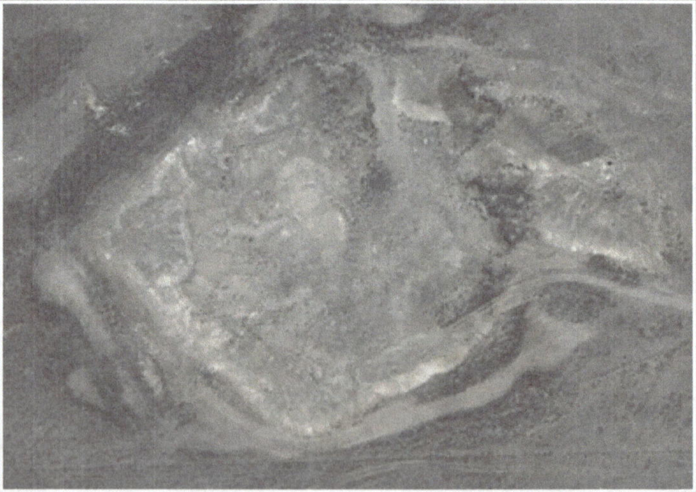

The mouths, plates, and the object angles are the exact design that pervades infinitely in a multiverse.

The Rosette Nebula (also known as Caldwell 49) is an H II region located near one end of a giant molecular cloud in the Monoceros region of the Milky Way Galaxy.
The nebula has been noted to be having a shape reminiscent of a human skull and is sometimes referred to as the "Skull Nebula.

The Earth specimen features are undeniably exact in every aspect to the angle of the eye sockets, encased skull and apex point, to the brow ridge definition.

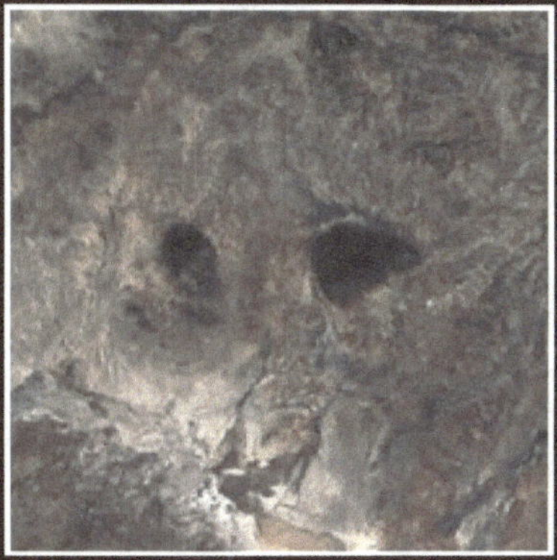

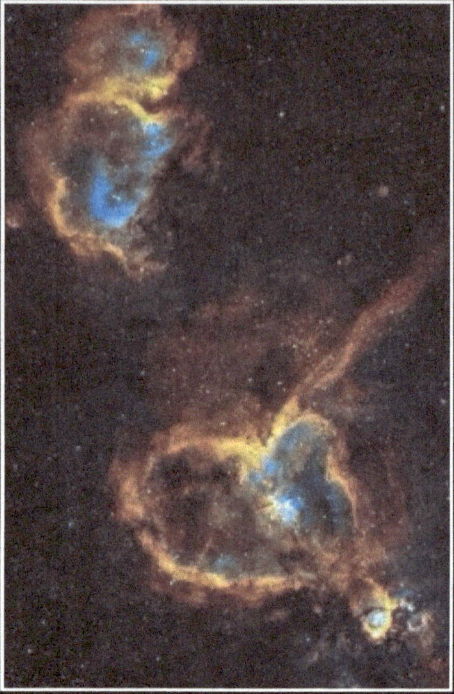

WESTERHOUT 5, (SHARPLESS 2-199, LBN 667, SOUL NEBULA) IS AN EMISSION NEBULA LOCATED IN CASSIOPEIA.

These Google Earth images (right) dramatically reveal the same superalloy processing as that of the galactic nebulae Soul,

7500 light years a way. The heart shape, eyes, curves, extending frontal tubes and head rods are stunningly coherent discovery features to the astro-geological functions.

Venus

A correlating astro-geological feature

From the galaxy to our solar system to parting of the Red Sea, this is major astro-geological evidence of the unfolding, replication on Earth and other planets.

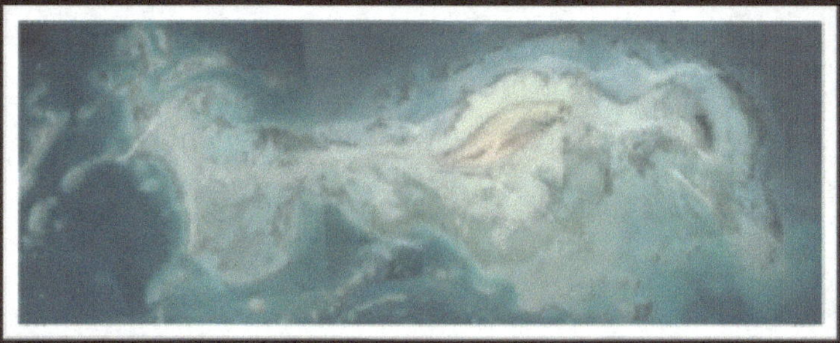

Assemblage, staging, replication and propulsion with limitless reactivation. Is this the inception of the Worm Hole in the Milky Way?
Google Earth images are45,000,000,000,000 mi. away (45 trillion).

The Mystic Mountain galaxy replicates to this 5 km. UT specimen. In all it's vast glory and mystery, a supreme life is revealed in form and functions which transcending beyond human space-time.

The oldest features on the ocean floor, are the best correlating datapoints to the cosmos.

JOHNNY TESONE

ABOUT THE AUTHOR

Johnny Tesone is a Colorado native, geologist, fly-fisherman, golfer, and former star athlete. A graduate of the University of Wyoming ('85, B.S. in Geology) on a full football scholarship, Johnny transitioned from the field to the frontier of geoscience after a career-ending concussion. He began his work as a mudlogger in the Rocky Mountain oil fields and went on to spend over four decades leading innovative geoscience and IT projects for major global firms.

Known for his sharp field observations and relentless curiosity, Johnny blends science with spiritual intuition. He believes he has been reincarnated to explore deep space and parallel Earths across dimensions of time.

His discoveries aim to empower the next generation to solve critical global challenges — harnessing fresh water, cultivating superfoods, and understanding the forces driving real climate change.